A Freshman Honors Course in

CALCULUS AND ANALYTIC GEOMETRY

By EMIL ARTIN

NOTES BY G. B. SELIGMAN

Foreword by Marvin J. Greenberg

Prepared under the auspices of the
Committee on the Undergraduate Program
Mathematical Association of America

1958

ISHI PRESS INTERNATIONAL

A Freshman Honors Course in Calculus and Analytic Geometry

by Emil Artin

First printed in 1957 in by the University of Virginia under the title
"A Freshman Honors Course in Calculus and Analytic Geometry
Taught at Princeton University"

Second Printing 1958 by the University of Buffalo under the same title

Third Printing 2008 by Ishi Press in New York and Tokyo

ISBN 0-923891-52-8
978-0-923891-52-7

Ishi Press International
1664 Davidson Avenue, Suite 1B
Bronx NY 10453
USA
917-507-7226

Printed in the United States of America

Publisher's Preface

Emil Artin, who was on born March 3, 1898, in Vienna, Austria and died on December 20, 1962, in Hamburg, Germany, was an Austrian mathematician. He was one of the great thinkers of his generation.

Although he was not Jewish, his wife was Jewish and for this reason he lost his job as a teacher in Germany in 1937. This proved to be fortunate in view of what happened later, because he came to America and in 1938 took a teaching position at Indiana University. He moved to Princeton University in 1946.

I became inspired to reprint this book as part of a new series of old math books because my wife is an accounting major and for some reason she must pass a course in basic calculus to graduate with a BS in Accounting. She took a course in calculus but discovered that the required textbook cost $216 !!!

She was having a hard time with calculus and so she dropped the course and signed up the next semester for the same course.

This time the required textbook was the same, but a new edition of the book had come out and using the elder edition of the same book was just not acceptable, so she had to spend another $216 for a new edition of the same book.

All this was covered by a student loan, but she will have to pay that back some day. I suppose that these student loans are what enables text book publishers to get away with the outrageous prices they are charging for these college textbooks.

This explains how today's students get into the endless cycle of ever increasing debt, which they hope to get out of with the big bucks they hope to make when (and if) they finally graduate.

Since there are a million old college math textbooks out there just as good if not better than the $216 basic math book my wife keeps having to buy, and since math has not changed much, I started looking for oldies but goodies in the math book field. I came across this book, entitled "*A Freshman Honors Course in Calculus and Analytic Geometry Taught at Princeton University*".

Publisher's Preface

While I realize that Princeton University was the number one math school in America, and for that reason the name is included in the title of the book, I have decided to reprint this book but omitting the part about "*Taught at Princeton University*" because other schools might want to adopt this book and naturally they think that they are just as good as Princeton University.

Fortunately, I discovered that an old friend, Dr. Marvin J. Greenberg, is a former student of Emil Artin. I know Dr. Greenberg from the time in the 1960s when I was a student at the University of California at Berkeley in the Math Department and Dr. Greenberg was an Assistant Professor there.

When I asked Dr. Greenberg to write a foreword to this book, I did not even know that he had been at Princeton University and that Emil Artin had been his teacher in 1956-1957, just before Artin had returned to Germany. This turned out to be a pleasant surprise.

This is the first in a series of old math books that Ishi Press will be reprinting. Next, we will be reprinting "*General Topology*" by John L. Kelley and "*Calculus with Analytic Geometry*" in two volumes by Angus E. Taylor.

Sam Sloan
Bronx NY
August 25, 2008

Foreword by Marvin J. Greenberg

Foreword

Emil Artin (1898 – 1962) was a leading algebraist of the twentieth century. He worked primarily in algebraic number theory, doing pioneering work in class field theory. He exhibited new L-functions generalizing the zeta functions, and (with his student John Tate) developed the cohomological approach to class field theory. He contributed importantly to the theories of groups, rings and fields. He invented the theory of braids in algebraic topology.

He gave a new, clearer exposition of Galois theory. The influential graduate text on abstract algebra by van der Waerden derived in part from Artin's ideas. He also wrote an elegant book on geometric algebra.

He had the rare distinction of having solved two of the famous problems posed by David Hilbert in 1900. He proved that every positive definite rational function of several variables was a sum of squares. He also discovered and proved the Artin reciprocity law, the grand culmination of over a century and a half of work on reciprocity laws. Artin had a great influence on the development of mathematics in his time, both by his many research contributions and by the high level of his teaching and expository writing.

He left two very fundamental conjectures. The first states that the Artin L-functions, known to be meromorphic, actually have no poles, with the possible exception of a pole at s = 1. The second asserts that a square-free integer a is a primitive root modulo infinitely many primes p, and the density of that set of primes is equal to Artin's constant, independent of a. A large body of research has been carried out on these conjectures, which remain unproven in their full strength so far, though significant partial results have been obtained. The vast Langlands Program provided a conjecture which implies Artin's first conjecture. The second conjecture has been shown to follow from the Generalized Riemann Hypothesis.

A third conjecture of Artin's about forms in many variables with coefficients in p-adic fields was shown by James Ax and Simon Kochen, for a given degree d, to be true for all but finitely many primes p (the possible exceptional primes depend on d). Research continues on finding more information about the exceptional primes, which have been shown to exist

in a few explicit cases.

A fourth conjecture of Artin about the absolute value of the roots of the zeta function of an elliptic curve over a finite field was proved by Helmut Hasse; the generalization of this result to arbitrary abelian varieties over a finite field was proved by André Weil, and Weil's further conjectures about zeta functions were eventually proved by Pierre Deligne using new cohomological methods developed by his teacher Alexandre Grothendieck. Thus Artin's conjecture led to enormously important results in arithmetic algebraic geometry.

A fifth Artin conjecture on approximating regular morphisms between Noetherian rings has been proved and has many applications in commutative algebra.

A sixth conjecture attributed jointly to Artin, Ankeny and Chowla asserts that if a prime p is congruent to 1 mod 4, then p does not divide the nth Bernoulli number, where n = (p-1)/2. This unproved conjecture has been verified by computer for all $p < 10^9$.

The quality of Artin's work is illustrated in the following quote by Richard Brauer. "There are a number of books and sets of lecture notes by Emil Artin. Each of them presents a novel approach. There are always new ideas and new results. It was a compulsion for him to present each argument in its purest form, to replace computation by conceptual arguments, to strip the theory of unnecessary ballast. What was the decisive point for him was to show the beauty of the subject to the reader."

Yet in his review of Bourbaki's Algèbre, Artin wrote: "We all believe that mathematics is an art. The author of a book or the lecturer in a classroom tries to convey the structural beauty of mathematics to his readers, to his listeners. In this attempt he must always fail. Mathematics is logical, to be sure; each conclusion is drawn from previously derived statements. Yet the whole of it, the real piece of art, is not linear; worse than that, its perception should be instantaneous. We have all experienced on some rare occasion the feeling of elation in realizing that we have enabled our listeners to see at a moment's glance the whole architecture and all its ramifications."

Artin supervised thirty four Ph.D. students, including the well-known

mathematicians Max Zorn, Hans Zassenhaus, Harold Shapiro, John Tate, Serge Lang, O.T. O'Meara, Bernard Dwork and Karel De Leeuw. His son Michael Artin is Professor of Mathematics at M.I.T., a leading expert in algebraic geometry.

Hans Zassenhaus wrote this about Artin's teaching: "The effect of listening to a lecture of Artin was to believe the subject on which he lectured was enormously simple, but upon subsequent reflection it became clear that the simplicity was the result of hard work and a life of dedication to pure mathematics ... Artin took a lively interest in teaching mathematics on all levels ... with wonderful clarity and patience he understood to guide his students to the full appreciation and mastery of the subtle concepts of limit and continuity, differentiability and integral from the beginning of their studies. His introductory course in analysis which I attended at the age of 17 converted me from a theoretical physicist to a mathematician."

Gian-Carlo Rota added this: "His lectures are best described as polished diamonds. They were delivered with the virtuoso's spontaneity that comes only after lengthy and excruciating rehearsal, always without notes. Very rarely did he make a mistake or forget a step in a proof... He would also snap back at students in the honors freshman calculus class which he frequently taught. He might throw a piece of chalk at a student who had asked too silly a question. A few weeks after the beginning of the fall term, only the bravest would dare ask any more questions, and the class listened in sepulchral silence to Emil Artin's spellbinding voice, like a congregation at a religious service."

As a first year graduate mathematics student at Princeton in 1955-6, I attended Artin's courses on Galois theory and introductory algebraic geometry. His lectures were models of clarity, and his gentle nobility was inspiring. As a result, I decided to do research in algebraic geometry for my Ph.D. thesis. But he was going on leave the following year, back to Hamburg, Germany where he was on the faculty before World War II. So he kindly arranged for his student Serge Lang, who was then Professor at Columbia University in New York, to commute to Princeton to teach an advanced course in algebraic geometry. I eventually wrote a thesis proving a conjecture of Lang's about abelian varieties over p-adic fields, using a new functor I invented.

Foreword by Marvin J. Greenberg

For an online biography of Emil Artin, see: http://www-history.mcs.st-and.ac.uk/Biographies/Artin.html .

Thus these notes from the honors freshman calculus course taught at Princeton by Artin are of considerable historical and pedagogical interest.

What makes the course specifically an "honors" course is best expressed in the following passage from G. B. Seligman's Introduction:

> **"Considerable stress is laid on active thinking in following the proofs. Perhaps the outstanding accomplishment of the course is imparting a sense of what constitutes a mathematical proof and an appreciation of the ingenuity which goes into the construction of a proof."**

But if that seems too much for a beginner: "Material is often first presented from a somewhat intuitive point of view." This provides preparation and motivation for the more rigorous presentation given later in this course.

So most of the material covered is standard for an honors first calculus course. Here are a few topics which are not found in most calculus texts:

p.75ff, the Approximation Theorem. Artin applies this to estimating the error in interpolating from tables. He also applies it to estimating the error in using Simpson's Rule for calculating a definite integral.

p.83. Artin derives the Lagrange remainder for Taylor's formula.

p.85. Artin reveals a trick for applying the Taylor series for arctan to effectively calculate many places in the decimal expansion of π .

p.88, the Bernstein Convergence Theorem for Taylor series. Artin shows cases in which this theorem can be applied. His important example is the binomial series on p.90.

pp.90-1. Artin gives an example where the Taylor series converges but not to

the original function.

pp.106ff. Artin provides axioms for the real number system. On pp.111-2 he shows how the existence of √2 can be proved from the least upper bound axiom. On pp.121-3 he also applies the least upper bound axiom to proving the two basic theorems about continuous functions on a closed interval.

p.114. Euler's constant is proved to exist as a limit.

p.119-20. Artin gives an example he calls "bizarre" of a function which is continuous at every irrational number but discontinuous at every rational number.

At the very end, it is mentioned that although the definite integral is only treated in terms of an intuitive notion of area in these notes, Artin did provide a rigorous discussion in the year following the one in which these notes were taken, and a reference to the book by Begle is given.

Marvin Jay Greenberg
Emeritus Professor of Mathematics
University of California

August 2008 in Berkeley

Emil Artin in 1927

Artin with Helmut Hasse

Emil Artin in 1927 in Hamburg, Germany at a dinner with some other famous mathematicians. Emil Artin is third from the left.

This photo from 1927 of many distinguished European mathematicians. Artin is third from the left, his official thesis advisor Gustav Herglotz fourth from the left, fifth from the right is Otto Schreier, with whom Artin collaborated on very important work, and first on the right is Bartel van der Waerden, who wrote that popular graduate algebra text influenced by Artin.

From left to right: Petersson, Furch, Artin, Herglotz, Reidemeister, Brauner, Haack, Hoheisel, Slotnik, Reinhardt, Schreier, Blaschke, Behnke, Kloosterman, van der Waerden

This photo which has no date but which must have been taken in the late 1920's shows Artin on the right when he was young, tall and handsome, and the great female algebraist Emmy Noether in the center. Unfortunately the man on the left is not identified.

A number of the students of Emil Artin became famous mathematicians. Here is a picture of Serge Lang, who edited "*The Collected Papers of Emil Artin*".

Here is a picture Marvin J. Greenberg with a visiting student. Dr. Greenberg is a former graduate student of Emil Artin and wrote the foreword to this book.

Dr. Greenberg was a graduate student of Emil Artin at Princeton University until Artin went back to Hamburg, Germany.

TABLE OF CONTENTS

THE HONORS COURSE IN FRESHMAN MATHEMATICS AT PRINCETON UNIVERSITY

INTRODUCTION

The average freshman enrollment in all mathematics courses at Princeton has, in recent years, been in the neighborhood of 550. From these an effort is made to select the students best qualified for the Honors course. The number selected has been reduced from about 75 in the early years of the course to about 60 in the year t954-55. A further reduction to about 50 seems advisable for the sake of quality. Selection is made on the basis of scores on the college board examinations, mathematical preparation, and interest in the program. When information as to the qualifications of a student is inconclusive, interviews with the departmental representative or with Professor Artin are used to obtain a further idea of his preparation and interest. Four years of secondary school mathematics is a customary requirement; this sometimes includes some contact with calculus, but no calculus or analytic geometry is prerequisite. A score of at least 700 on the advanced mathematics test of the college board examinations is usually required, but is not always a sufficient condition for admission to the course. Interest in the program among the students seems to be no problem; in fact, it is necessary to disappoint a number of applicants annually. These are usually enrolled in advanced sections of the standard course in analytic geometry and calculus.

At its beginning, the course was presented in three or four sections, each of which met three times weekly. One of these sections was taught by Professor Artin, the others by full-time faculty members. Under this arrangement, the absence of unified textual material contributed to less coordination among the sections than might be desired, and variations in grading procedures were sometimes harmful to morale among the students. In 1954-55, with a slightly smaller group than before, the presentation was changed. Professor Artin gave two lectures weekly to the entire group, which was divided into five sections of about twelve students each for the third hour of the week. This procedure is to be continued in the coming year.

Due to the interest of the students and the fact that their questions are encouraged, the lectures are conducted with close contact between Professor Artin and the individual students: He continually solicits their participation in answering questions, in carrying out proofs and in the formulation of new concepts (the derivative is a notable example). Considerable stress is laid on active thinking in following the proofs. Perhaps the outstanding accomplishment of the course is imparting a sense of what constitutes a mathematical proof and an appreciation of the ingenuity which goes into the construction of a proof. The fact that this has been achieved to any considerable extent seems to more than justify the existence of the program.

In the past year, all five recitation sections were conducted by the same instructor. At their meetings a question-and-answer session was very common; homework problems and other exercises were worked; material from the lectures was discussed; and new material of a rather specialized nature was occasionally presented. Of the outline of Calculus which follows, Sections 9, 13, 22 and the last part of Section 24 were the only portions presented independently in these classes.

The careful screening of students admitted to the course enables us to assume that, with a reasonable amount of effort, the students in the Honors course would stand at and near the head of their class in the standard course. They are not penalized in terms of grades for volunteering for the Honors course; a student whose classroom participation and homework have indicated serious effort is assured of a high grade. This is further guaranteed by giving examinations which test for knowledge of the rudiments, rather than giving specialized problems and techniques involving extensive calculation. These procedures no doubt contribute to the popularity of the course and to the student attitude that admission is a privilege. The students are made aware that it is their responsibility, in return for this privilege, to apply themselves diligently without the prodding provided by frequent testing. (Four examinations a year is the rule.) They are held responsible for all material presented. Obvious laggards may be "demoted" to the standard course.

The course program which follows is essentially that

which was used in 195-55. Certain omissions from the customary course will be noted; tonics, polar coordinates, the "subdivision" approach to the integral and its existence for continuous functions, physical applications - to name several. Some topics treated are seldom to be found in a freshman course. The order of presentation is such that material is often first presented from a somewhat intuitive point of view. This achieves the twofold result of making it seem less formidable to the student and of giving him preparation and motivation for the more rigorous presentation later in the course. Since there is a sophomore honors course which constitutes a natural sequel to that for freshmen and which is taken by most of the same students, one need not be seriously concerned about the areas in which the rigorous presentation does not catch up with the intuitive one; they will be covered. The question of physical applications is a more pressing one, since many of the students are in engineering and physics. It has been suggested (by the student engineering council) that optional collateral reading and problems be supplied to make up this deficiency, and this suggestion may be adopted in the coming year.

Even with superior students, the difficulty of getting them to substitute is a serious one, and is possibly the greatest handicap to progress. For instance, if one shows that $A^2 \geqq 0$ if A is any vector, $(B + C)^2 \geqq 0$ causes many students some concern. Considerable repetition of emphasis on the principle of substitution, punctuated by such outlandish substitutions as "bird" and "crocodile" for the object in question, eventually reduces this obstacle.

Apart from teaching an appreciation of mathematical proofs, much of the unique merit of the course lies in aspects not to be found in the formal program. Frequently points in the lectures stimulate discussion of unplanned material. This material may range from the meaning of negative numbers to the distinction between real and cardinal numbers and questions of countability. In addition, entire periods 'are often devoted to discussion of random mathematical questions proposed by the students. In the past year these have included a question concerning the problems of knot theory and a discussion of "life on a torus". In connection with the n-dimensional approach to analytic geometry, it was found of interest to the student to try to liberate him from the 3-dimensional viewpoint by asking him to visualize various configurations in four-space with the three ordinary spatial coordinates and one time coordinate. These digressions are not regarded as incidental, but rather as an opportunity for the student to increase his awareness of the scope and nature of mathematics.

ANALYTIC GEOMETRY

1. Points

In high school, one learns to think of a point as something which has no size but to which one can assign a definite location. But this concept is so idealized that we can not realize it in nature. For certainly any dot we draw on the blackboard has dimensions, and the principles of modern physics tell us that the determination of a precise location for an object is impossible. Therefore the things which we shall call points must come from an ideal world, and that will be the world of numbers.

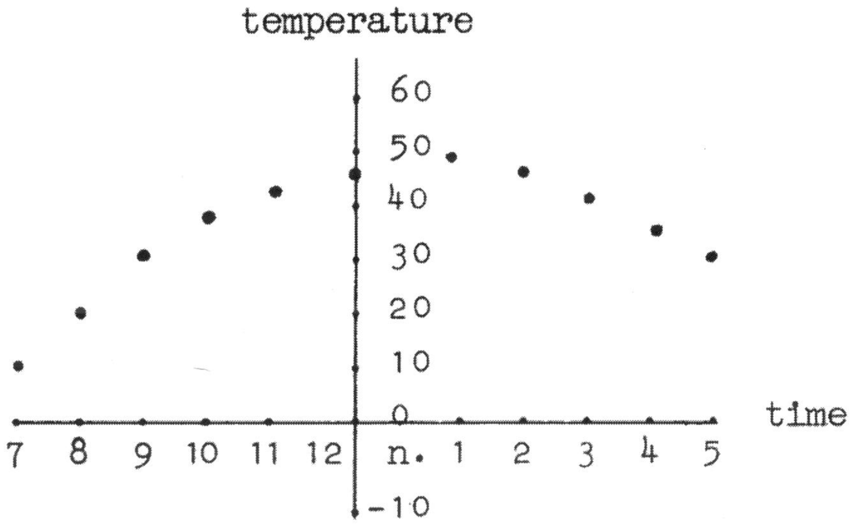

Each of the dots we have indicated, or each point of our graph, corresponds to a pair of numbers, e.g., (9 A.M., 30°). The first number indicates the (signed) distance from a fixed vertical line of reference; this point is - 3 hours from the noon-line. Similarly the second number indicates the distance above the zero-degree line, in this case + 30 degrees.

If we could extend indefinitely the range of available times and temperatures, it is evident that we could represent every point in the plane of the paper by a pair of numbers, and by only one such pair. For instance a point well off to the right and well below the zero-degree line might be represented by (noon two days later, - 379°), or simply (48, - 379). It is seen, then, that once a pair of reference lines (or axes) has been chosen, we can consider a point in the plane simply as a pair of numbers, representing its distances from the reference lines.

Likewise in space, we take three lines which intersect in a point. They then determine three planes, and we can identify every point in space with the triple of numbers consisting of its signed distances from these planes. Proceeding from this motivation, we define a point in n-dimensional space to be an n-tuple

$(x_1, x_2, \ldots, x_n)$ of numbers.

Then n-dimensional space is defined to be the totality, or set, of all
n-tuples $(x_1, x_2, \ldots, x_n)$ of numbers.

1. Distance

Now we return to two-dimensional space, where the notion of distance is a familiar one. In the figure below, (x_1, x_2) and (y_1, y_2)

are two points in the plane, and the distance between them is the hypotenuse of the triangle with vertices

$$(x_1, x_2), (y_1, x_2) \text{ and } (y_1, y_2),$$

or

$$\sqrt{(y_1 - x_1)^2 + (y_2 - x_2)^2}$$

by the theorem of Pythagoras.

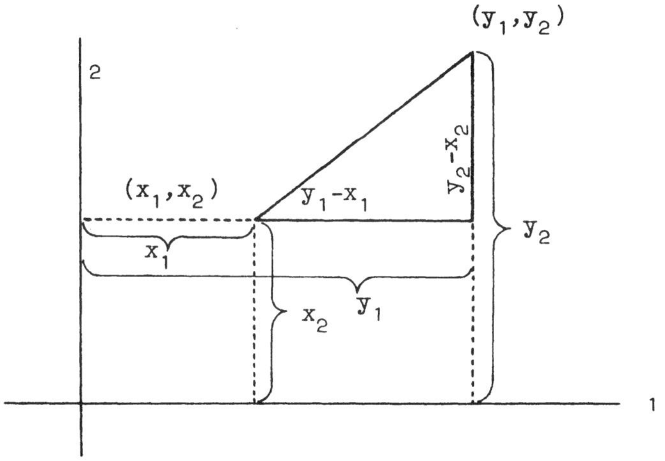

It is of some interest to give a very simple and often unfamiliar proof of the theorem of Pythagoras, as follows: We wish to prove that if a and b are the legs of a right triangle and c is its hypotenuse, then $a^2 + b^2 = c^2$. Let us form a square of side $a + b$, and decompose it in two ways:

(1)

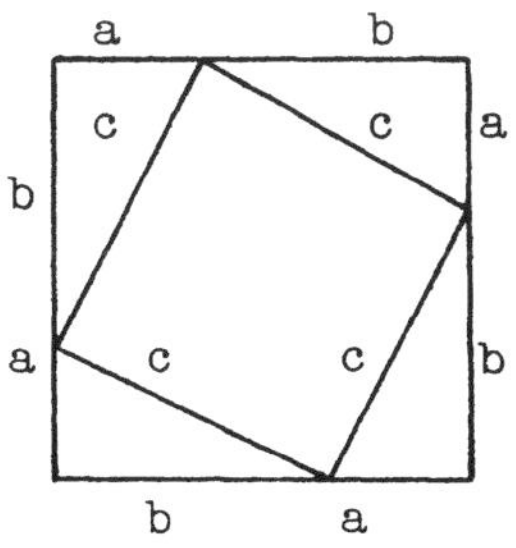

(2)

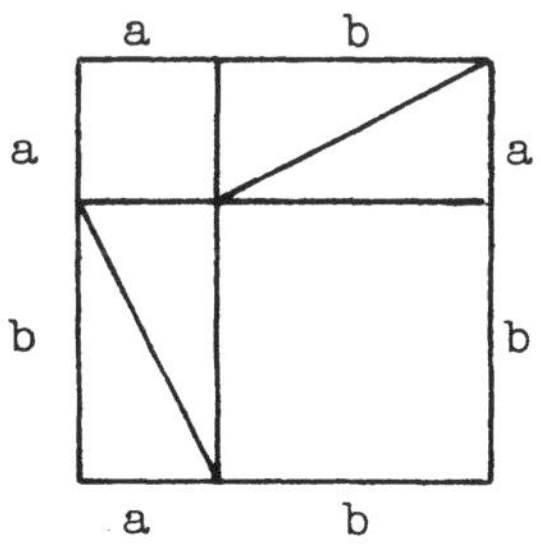

From (1), the area of the square is

$$c^2 + 4\Delta \ ,$$

where Δ is the area of a right triangle with legs a and b. From (2), its area is

$$a^2 + b^2 + 4\Delta \ .$$

Thus

$$a^2 + b^2 + 4\Delta = c^2 + 4\Delta \ ,$$

or

$$a^2 + b^2 = c^2 \ ,$$

which was to be proved.

By applying the theorem of Pythagoras twice, we see that the distance between two points (x_1, x_2, x_3) and (y_1, y_2, y_3) in 3-space is

$$\sqrt{(y_1 - x_1)^2 + (y_2 - x_2)^2 + (y_3 - x_3)^2} \ .$$

Thus we are led to define the *distance between two points* $(x_1, x_2, \ldots, x_n)$ and $(y_1, y_2, \ldots, y_n)$ in n-space to be

$$\sqrt{(y_1 - x_1)^2 + (y_2 - x_2)^2 + \ldots + (y_n - x_n)^2} \ .$$

We observe that this is equal to

$$\sqrt{(x_1 - y_1)^2 + (x_2 - y_2)^2 + \ldots + (x_n - y_n)^2} \ .$$

We also note that the distance between two points is 0 only if the points are not actually distinct, i.e., only if $x_1 = y_1, x_2 = y_2, \ldots, x_n = y_n$. (Why?)

NOTE. When we write $b = \sqrt{a}$, where a is positive or zero, we refer to that number b which is positive or zero and such that $b^2 = a$.

3. *Vectors*

In physics, one often refers to a *vector* as a quantity with length and direction, and represents it by an arrow ↗ . This can hardly be called a definition, since almost all the terms used are *undefined*. To make it precise, observe that the "arrow" is determined if we specify:

a) its initial point (the tail of the arrow);
b) its end point (the head of the arrow).

Thus we are led to define a *vector in n-space* to be an *ordered pair* of points in n-space, therefore an ordered pair of n-tuples of numbers. The first point will be called the <u>initial point</u>, the second the <u>end point</u>.

It is likewise customary to regard two vectors as <u>equal</u> if they have the same length and the same direction, thus if their "arrows" satisfy these conditions. Let us look at this in 2-space:

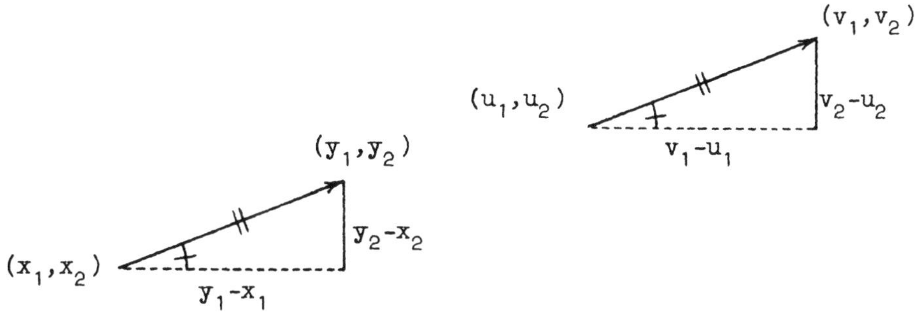

If these two vectors are to be equal in this sense, it means that the two right triangles shown have their hypotenuse and one acute angle equal, therefore are congruent. In particular, the other sides are equal, i.e.,

$$y_1 - x_1 = v_1 - u_1, \quad y_2 - x_2 = v_2 - u_2 \quad .$$

Therefore let

$$((x_1, x_2, \ldots, x_n), \quad (y_1, y_2, \ldots, y_n))$$

and

$$((u_1, u_2, \ldots, u_n), \quad (v_1, v_2, \ldots, v_n))$$

be two vectors. We define them to be <u>equal</u> if $y_1 - x_1 = v_1 - u_1$, $y_2 - x_2 = v_2 - u_2, \ldots, y_n - x_n = v_n - u_n$.

It is seen, then, that a vector is determined to within equality by the numbers $a_1 = y_1 - x_1$, $a_2 = y_2 - x_2, \ldots, a_n = y_n - x_n$. Thus one is led to consider a quantity which one writes as $[a_1, a_2, \ldots, a_n]$. The numbers $a_1, a_2, \ldots, a_n$ are called <u>components</u> of the vector, and the expression $A = [a_1, a_2, \ldots, a_n]$ is called the <u>free vector</u> determined by the vector

$$((x_1, x_2, \ldots, x_n), \quad (y_1, y_2, \ldots, y_n))$$

(or by any vector equal to this one).

Now we see how to find the free vector determined by any (ordered pair of points, and also how to find, for a given free vector and a given point, a second point such that the vector with the first point as initial point and the second as end point determines the given free vector. Given $A = [a_1, a_2, \ldots, a_n]$ and initial point $(x_1, x_2, \ldots, x_n)$, the desired end point is of course $(x_1 + a_1, x_2 + a_2, \ldots, x_n + a_n)$. We shall often use the term located vector to refer to a pair of points, and the term "vector" will rapidly be usurped by the free vector.

The principle of addition of vectors is familiar from the parallelograms of forces used in physics. For the moment, let $A = [a_1, a_2]$ and $B = [b_1, b_2]$ be two vectors in 2-space. Let A be located with initial point (x_1, x_2), end point (y_1, y_2). Then we have $y_1 = x_1 + a_1$, $y_2 = x_2 + a_2$. We form $A + B$ by locating B with initial point (y_1, y_2). Let (z_1, z_2) be the end point of B. Then $A + B$ is the vector determined by the pair of points $((x_1, x_2), (z_1, z_2))$.

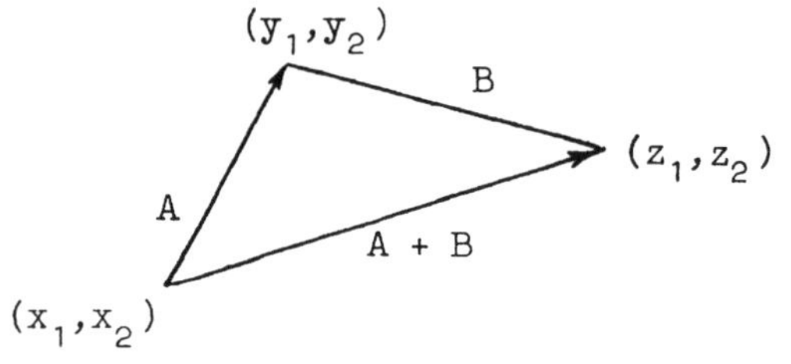

But $z_1 = y_1 + b_1 = x_1 + a_1 + b_1$, $z_2 = y_2 + b_2 = x_2 + a_2 + b_2$. Thus we have, for the free vectors, $A + B = [a_1 + b_1, a_2 + b_2]$.

This leads us to define, for arbitrary n,

$$[a_1, a_2, \ldots, a_n] + [b_1, b_2, \ldots, b_n] = [a_1 + b_1, a_2 + b_2, \ldots, a_n + b_n].$$

Please note that this is a definition, so that it requires no proof. The things that do require proof are the elementary properties of this operation, and they are given as exercises:

i) $A + B = B + A$

ii) $(A + B) + C = A + (B + C)$

iii) There is a vector 0 such that for any vector A, $A + 0 = A$. (How many are there?)

iv) Given any vector A, there is a vector $(- A)$ such that $A + (- A) = 0$, where 0 is as in iii). (How many are there?)

(NOTE. It seems advantageous at this stage to introduce the student to the words associative and commutative, and to the concept of a group. An opportunity is provided to present simple examples of operations which are non-commutative or which fail to satisfy some of the group axioms.)

Let $A = [a_1, a_2, \ldots, a_n]$ be a vector in n-space. We define the length of A to be the distance between its initial point and its end point when it is located at some point. Let A be located with initial point $(x_1, x_2, \ldots, x_n)$, end point $(y_1, y_2, \ldots, y_n)$. Then

$y_1 = x_1 + a_1$, $y_2 = x_2 + a_2$, ..., $y_n = x_n + a_n$. The length of A is thus

$$\sqrt{(y_1 - x_1)^2 + (y_2 - x_2)^2 + \dots + (y_n - x_n)^2}$$

$$= \sqrt{a_1^2 + a_2^2 + \dots + a_n^2} \ .$$

From this we see that the point at which we located the vector A does not affect its length. We shall denote this <u>number</u> (<u>not</u> a vector) by $|A|$. It is sometimes called the <u>absolute value</u> of A. We thus have the rule

$$|A| = \sqrt{a_1^2 + a_2^2 + \dots + a_n^2} \ .$$

We now have means of calculating with two categories of objects; vectors and numbers (or <u>scalars</u>). We shall introduce more operations, which will mix these two categories. It should, however, be remembered that the two are distinct, and that an equation $A = 2$, for instance, is as meaningless as the equation "x sheep = 2 oranges". Due to convention and the lack of suitable symbols, we use 0 to represent both the number zero and the zero-vector of iii). No confusion should arise from equations involving this symbol, however, because the rest of the equation will enable us to interpret the category correctly.

Next we define <u>multiplication of vectors by numbers</u>; if $A = [a_1, a_2, \dots, a_n]$ and b is a number, we define

$$bA = [ba_1, ba_2, \dots, ba_n] \ .$$

Note, for example, that for $b = 3$, we have $3A = A + A + A$, as in the case of numbers. Again some properties are given as exercises:

v) $(a + b)C = aC + bC$

vi) $a(B + C) = aB + aC$

vii) $0A = 0$

viii) $1A = A$, $(-1)A = -A$

ix) $a(bC) = (ab)C$.

Consider two (located) vectors $((x_1, x_2), (y_1, y_2))$ and $((u_1, u_2), (v_1, v_2))$ in 2-space. When do we say that they have the same direction? In the following diagram this amounts to saying that the two angles indicated by $\sphericalangle$ are equal; that is, they are equal in magnitude and also in sign, where they are measured counterclockwise from the positive x-axis (i.e., the horizontal axis). Thus the <u>tangents</u> of these angles are equal, or

$$\frac{y_2-x_2}{y_1-x_1} = \frac{v_2-u_2}{v_1-u_1} .$$

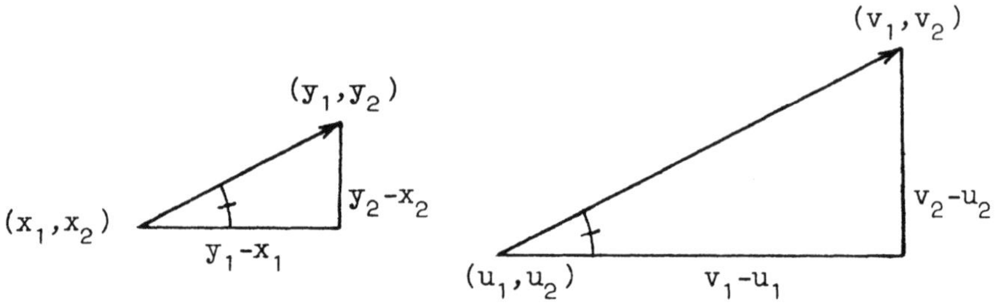

Multiplying by

$$\frac{y_1-x_1}{v_2-u_2}$$

gives

$$\frac{y_2-x_2}{v_2-u_2} = \frac{y_1-x_1}{v_1-u_1} .$$

Let us call this number b. Then $[y_1 - x_1, y_2 - x_2] = b[v_1 - u_1, v_2 - u_2]$. Moreover, $b > 0$ (i.e., b is positive), as is evident from the diagram. (We have assumed in performing divisions that the vectors are neither vertical nor horizontal; it is left as an exercise to show that the same type of result holds in these cases.) Thus we are led to make the following definitions:

Let A, B be non-zero vectors in n-space. We say that A and B have the <u>same direction</u> if $B = bA$ for some number $b > 0$. We say that A and B have <u>opposite directions</u> if $B = bA$ for some number $b < 0$. We say B is <u>parallel to</u> A if A and B either have the same direction or have opposite directions, therefore if $B = bA$ for some number $b \neq 0$. (Note that then A is also parallel to B.)

4. Straight Lines

We have written two types of n-tuples of numbers: the n-tuple $(x_1, x_2, \ldots, x_n)$ is a point in n-dimensional space, while the n-tuple $A = [a_1, a_2, \ldots, a_n]$ is a free vector in n-space. We now drop the notation of the square brackets and write $A = (a_1, a_2, \ldots, a_n)$. The apparent confusion resulting between points and free vectors is eliminated as follows. Given the point $(x_1, x_2, \ldots, x_n)$, consider the free vector determined by the located vector with initial point $(0, 0, \ldots, 0)$ and end point $(x_1, x_2, \ldots, x_n)$. In our new notation, this vector is also $(x_1, x_2, \ldots, x_n)$. Thus one may consider the n-tuple $(x_1, x_2, \ldots, x_n)$ as representing a point in n-space, or as representing the vector with $(0, 0, \ldots, 0)$ as initial point and this point as end point. By this identification, we may use vector addition to add objects which are ostensibly points, and then identify the resulting vector with a point.

Now we shall define a line in n-space. This is an object of a different category, neither point nor vector nor number. It is of the same category as the whole of n-dimensional space, in that it is a set, or collection of points. Thus to determine a line, we must give a rule which will enable us to decide whether a given point is on that line, or as we say, is a member of that set of points.

It is a familiar property of plane geometry that a line is determined by giving a point on it and a direction (say another line, or a vector) parallel to it. It is of course also determined by giving two distinct points on it, but this property is less convenient for our purposes. Let $P = (p_1, p_2)$ be a point in 2-space, and let $A = (a_1, a_2)$ be a vector, $A \neq 0$. The question we must then answer is: Given a point $X = (x_1, x_2)$, when is it on the line through P and parallel to A? It is clear that the line segment $\overline{PX}$, or, rather, the vector with initial point P and end point X, must be parallel to A.

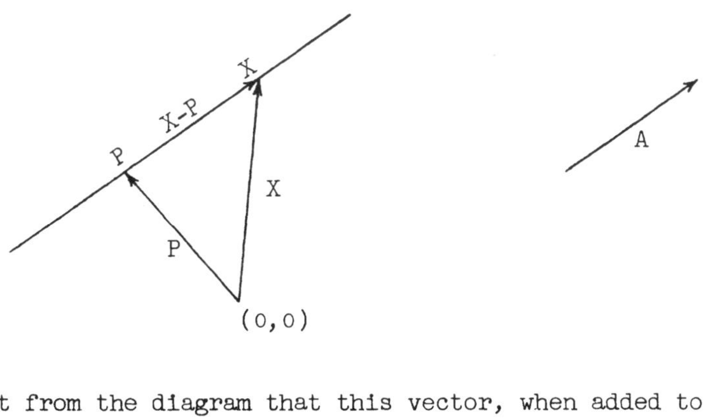

It is evident from the diagram that this vector, when added to the vector P (using our identification of points and vectors), gives the vector X. Hence the vector with initial point P and end point X is the vector

$X - P$ (meaning $X + (-P)$; see the exercises on vector addition). Thus we must have $X - P = tA$, where t is some non-zero number. Of course, if we set $t = 0$, then $X = P$ is on the line. Thus $X = P + tA$, where t is some number. Moreover, if t is any number, the point $X = P + tA$ has the property that either $X = P$ or $X - P$ is parallel to A, i.e., X lies on the desired line.

Now let P be a point in n-space, A a (non-zero) vector in n-space. We define the <u>line through</u> P <u>parallel to</u> A to be <u>the set of all points</u> X <u>such that</u>

$$X = P + tA$$

<u>for some number</u> t, i.e., where t <u>ranges</u> through all numbers.

In 2-space, this takes the form $(x_1, x_2) = (p_1 + ta_1, p_2 + ta_2)$. Since $A \neq 0$, either $a_1 \neq 0$ or $a_2 \neq 0$. Let us assume $a_1 \neq 0$. Then

$$x_1 = p_1 + ta_1, \quad \text{or} \quad t = \frac{x_1 - p_1}{a_1},$$

and

$$x_2 = p_2 + ta_2 = p_2 + \left(\frac{x_1 - p_1}{a_1}\right) a_2 = \frac{a_2}{a_1} x_1 + \left(p_2 - p_1 \cdot \frac{a_2}{a_1}\right).$$

If we change our notation and write x for x_1, y for x_2, m for $\frac{a_2}{a_1}$ and b for $p_2 - p_1 \frac{a_2}{a_1}$, we obtain

$$y = mx + b,$$

which is the ordinary "slope-intercept" equation of a line. It means that the point (x, y) is on the line if and only if $y = mx + b$, where m and b are the numbers defined above. The form $P + tA$ for the equation of a line is customarily called the <u>parametric form</u>. It is used in this course because of its complete generality for all dimensions and for all vectors $A \neq 0$.

Now let $P = (p_1, p_2, \ldots, p_n)$ and $Q = (q_1, q_2, \ldots, q_n)$ be two points. What is the line through P and Q? We have a point, say P, on the line, and a vector parallel to the line, namely the vector determined by the initial point P and end point Q. But we have seen that this vector is just $Q - P$. Thus the desired line is the set of points

$$X = P + t(Q - P)$$

where t ranges through all numbers, or the set of points

$$X = (1 - t)P + tQ$$

where t ranges through all numbers.

5. Scalar Products

The next definition we shall make is that of scalar product, whereby two vectors are combined to give a number. Let $A = (a_1, a_2, \ldots, a_n)$, $B = (b_1, b_2, \ldots, b_n)$. We define the scalar product AB by

$$AB = a_1b_1 + a_2b_2 + \ldots + a_nb_n .$$

It is left as an exercise to prove the following:

i) $A(B + C) = AB + AC$.
ii) $AB = BA$.
iii) $a(BC) = (aB)C = B(aC)$.

In particular, we define $A^2 = AA$. Thus $A^2 = a_1^2 + a_2^2 + \ldots + a_n^2$. But if we recall that

$$|A| = \sqrt{a_1^2 + a_2^2 + \ldots + a_n^2} ,$$

we have

$$A^2 = |A|^2 .$$

(Note that this equation makes sense, since both sides are numbers. Note also that if $A \neq (0, 0, \ldots, 0)$, $A^2 > 0$.)

Now let us try to express, in terms of the language we have developed, what is to be meant by saying that two vectors are perpendicular. Suppose that A and B of the diagram are perpendicular. In particular, this means that if we lay off a segment of length $|B|$ to the left of their common initial point, the two line segments marked $||$ are of equal length. But if we consider the broken line segment as a vector directed to the left, we observe that this vector is $-B$, and that the two equal line segments, directed as indicated by the arrows, are $A - B$ and $A - (-B) = A + B$. Thus a necessary condition for perpendicularity is $|A - B| = |A + B|$.

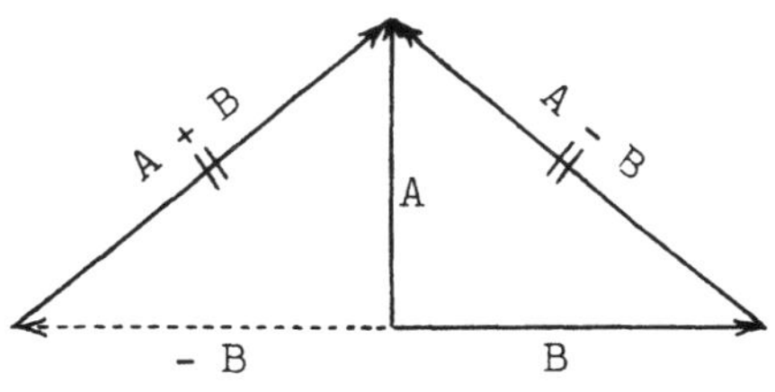

Conversely, if this is the case, then A is perpendicular to B in the sense of plane geometry. Squaring both sides of this equation and using the fact that $A^2 = |A|^2$ for <u>any</u> vector A, we have

$$(A - B)^2 = |A - B|^2 = |A + B|^2 = (A + B)^2 .$$

Using the rules developed in the exercises,

$$(A - B)^2 = A^2 - 2AB + B^2, \quad (A + B)^2 = A^2 + 2AB + B^2.$$

Thus $A^2 - 2AB + B^2 = A^2 + 2AB + B^2$, or

$$-2AB = 2AB, \quad \text{i.e.,} \quad \underline{AB = 0} .$$

Conversely, if $AB = 0$, we see that $|A - B|^2 = |A + B|^2$. But since both $|A - B|$ and $|A + B|$ are non-negative, this means that $|A - B| = |A + B|$, or that A and B are perpendicular.

Thus we may regard this result as our definition of perpendicularity in n dimensions: A <u>is perpendicular to</u> B will mean simply that $AB = 0$.

Now suppose A and B are non-zero vectors (recalling that the zero vector is $(0, 0, \ldots, 0)$). Then the quantity

$$|B|A - |A|B$$

is a vector. Therefore the <u>number</u> $(|B|A - |A|B)^2$ is either positive or zero, and is zero only if $|B|A - |A|B = 0$ (vector), i.e., only if

$$|B|A = |A|B.$$

Now

$$\begin{aligned}(|B|A - |A|B)^2 &= |B|^2A^2 - 2|A|\,|B|AB + |A|^2B^2 \\ &= |B|^2|A|^2 - 2|A|\,|B|AB + |A|^2|B|^2 \\ &= 2|A|\,|B|(|A|\,|B| - AB) ,\end{aligned}$$

and this is $\geqq 0$. Since $2|A|\,|B|$ is positive, we can cancel it and get $|A|\,|B| - AB \geqq 0$, or

$$AB \leqq |A|\,|B| .$$

The two are equal only if $|B|A = |A|B$, i.e., only if A and B have the same direction.

Likewise, $(|B|A + |A|B)^2 \geqq 0$, and is zero only if

$|B|A = -|A|B$, i.e., only if A and B have opposite directions. But

$$(|B|A + |A|B)^2 = 2|A|\ |B|(|A|\ |B| + AB) \geqq 0,$$

giving

$$|A|\ |B| + AB \geqq 0, \quad \text{or} \quad -|A|\ |B| \leqq AB.$$

Equality holds only if A and B have opposite directions. Combining the two results, we have in general

$$-|A|\ |B| \leqq AB \leqq |A|\ |B|,$$

and we know exactly when one or the other of the inequalities becomes an equality. This result is known as Schwarz's inequality.

Dividing the inequality through by $|A|\ |B|$ (which is positive), we have

$$-1 \leqq \frac{AB}{|A|\ |B|} \leqq 1.$$

Therefore the number $\frac{AB}{|A|\ |B|}$ can be the cosine of some angle. If we require that this angle θ be between $0°$ and $180°$ (or in radians, between 0 and π), then θ is uniquely determined by giving $\cos\theta$. Thus we define the angle between A and B to be that angle θ between $0°$ and $180°$ such that

$$\cos\theta = \frac{AB}{|A|\ |B|}.$$

Observe that $\theta = 0°$ if and only if A and B have the same direction, $\theta = 180°$ if and only if they have opposite directions. So our terminology is consistent in this respect. Plane trigonometry will provide a more complete justification of our choice of definitions. Consider the triangle shown. The law of cosines tells us that

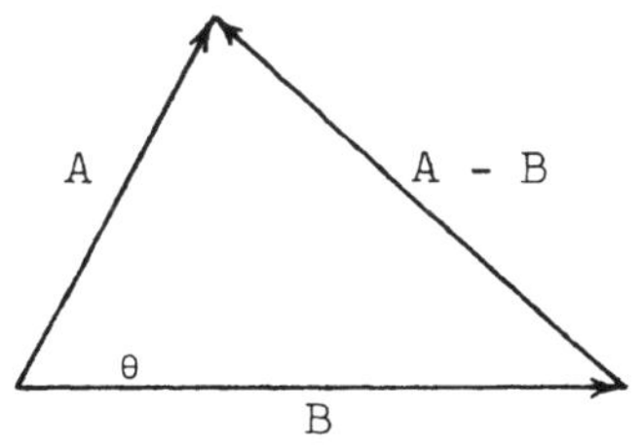

$$|A - B|^2 = |A|^2 + |B|^2 - 2|A|\ |B| \cos\theta.$$

Using what we know about vectors, this means that

$$(A - B)^2 = A^2 + B^2 - 2|A|\ |B| \cos\theta,$$

or

$$A^2 - 2AB + B^2$$

$$= A^2 + B^2 - 2|A|\ |B| \cos\theta,$$

i.e.,

$$-2AB = -2|A|\ |B| \cos\theta,$$

and finally

$$AB = |A|\ |B| \cos\theta,$$

which agrees with our definition of θ.

6. Planes

We shall next define another geometrical configuration in n-space; its category is again that of a set of points. It will be called a plane, and once again it is determined by giving a point and a vector. Our experience in geometry tells us that in ordinary 3-space a plane is completely determined if we know a point on the plane and a direction (i.e., a vector or a line segment) perpendicular to the plane. We shall find an equation which will provide a rule for deciding whether a given point is on our plane. For a given point P and a given vector A, the set of points whose coordinates satisfy this equation will then be called the plane through P and perpendicular to A.

Suppose $X = (x_1, x_2, \ldots, x_n)$ is any point on such a plane. Then the line segment joining X and P must lie in this plane and therefore must itself be perpendicular to A. Considering this segment as a vector X - P, as indicated in the diagram, we see that $(X - P)A = 0$, or that

$$\underline{AX=AP}\ .$$

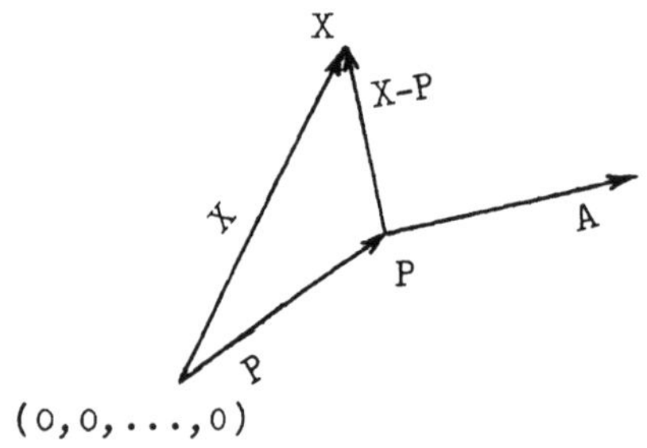

This is then the desired equation, and the plane is the set of all points $X = (x_1, x_2, \ldots, x_n)$ satisfying this relation.

For example, the equation of a plane in 3-space would look like

$$4x - 7y + 3z = 13,$$

where we use $X = (x, y, z)$ instead of (x_1, x_2, x_3). This is of the form $AX = AP$, where $A = (4, -7, 3)$. The point $(1, 0, 3)$ is a point of the plane, which we could take as our P. We see from this how to recover a vector perpendicular to a plane from the equation: we simply form a vector whose components are the coefficients of x, y and z, in this notation.

It is also evident how to determine from a pair of equations of this form whether the planes they define are parallel. For this question is equivalent to the question of parallelism for two vectors perpendicular to the planes. Since such vectors are given by reading off the coefficients of x, y and z, we see that the two planes are parallel if and only if these coefficients are proportional.

7. Vector Product

Because of its applications to physics, we shall define still another operation with vectors, which is again called a product. This product is only defined in case the vectors to be multiplied are vectors in 3-space, and its category is that of a vector. For the latter reason, it is called the vector product. It is written $A \times B$ and is read "A cross B".

Let $A = (a_1, a_2, a_3)$, $B = (b_1, b_2, b_3)$. Then we define

$$A \times B = (a_2b_3 - a_3b_2, a_3b_1 - a_1b_3, a_1b_2 - a_2b_1).$$

Exercises:

i) $A \times B = -(B \times A)$.

ii) $A \times (B + C) = (A \times B) + (A \times C)$.

iii) $(aA) \times B = a(A \times B) = A \times (aB)$

iv) $(A \times B) \times C = (AC)B - (BC)A$

v) $((A \times B) \times C) + ((B \times C) \times A) + ((C \times A) \times B) = 0$

vi) $A \times B$ is perpendicular to both A and B.

vii) $(A \times B)^2 = A^2B^2 - (AB)^2$. Use this to give a geometric meaning to $|A \times B|$.

8. Applications and Problems

We shall present a few applications of the technique of vectors to geometry, followed by some problems of more than routine interest. This list should be supplemented by a number of the problems to be found in any of the standard texts on analytic geometry of two and three dimensions.

1. Consider the problem of finding the intersection of two planes in 3-space. In general, this will be a line; we can derive its equation from those of the planes in several ways:

a) We could solve simultaneously the equations of the planes.

b) We could find two points which are common to both planes, for instance by choosing two different fixed values for their third coordinates. We then obtain in each case a system of two linear equations in two unknowns (the first and second coordinates), which we solve. Then we find the connecting line.

c) We could find one point P common to both planes, and then find a vector parallel to the line of intersection. The latter is accomplished by observing that any such vector A must be perpendicular to each of the known vectors B, C perpendicular to the individual planes. From the exercises on vector products, we see that we can take as our vector A the vector product $B \times C$. The formula $P + tA$ gives the line.

2. Let $P + tA$ be a line in n-space, Q a point not on the line; let us find the foot of the perpendicular from Q to this line.

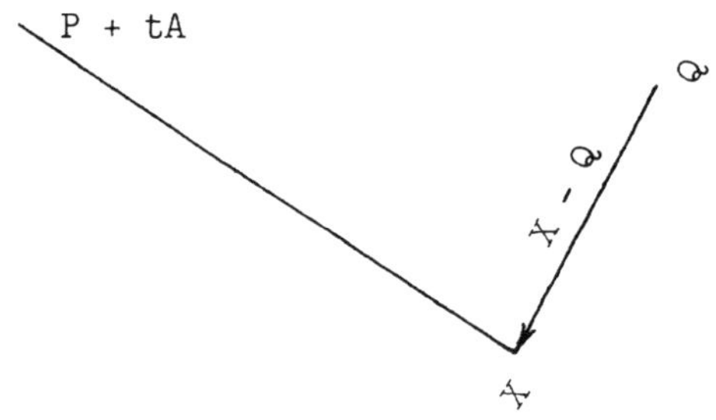

If X is this point, then $X = P + t_0A$ for some t_0. Moreover, the segment connecting X with Q, or the vector $X - Q$, is to be perpendicular to the line $P + tA$, therefore to the vector A. Now $X - Q = P + t_0A - Q$, giving

$$(P + t_0A - Q)A = 0.$$

This is a linear equation, which we solve for t_0. The desired point is then found by substituting this value for t_0 in $X = P + t_0A$. The distance between Q and the line is simply $|X - Q|$.

3. Now let $P + tA$ and $Q + sB$ be two non-intersecting lines. Let $X = P + t_0A$ and $Y = Q + s_0B$ be the endpoints of their common perpendicular. Then $(X - Y)A = 0$ and $(X - Y)B = 0$. These give us two linear equations which we solve for s_0 and t_0. The resulting points X and Y determine the common perpendicular. The distance between the two lines is $|X - Y|$.

4. The angle between two planes is found by using the angle between two vectors, i.e., by reading off two vectors perpendicular to the planes from their equations and finding the angle between these vectors. Since two vectors having opposite directions determine the same plane,

there are two possible answers, consisting of supplementary angles. We agree to take the angle between 0° and 90°.

PROBLEMS

1) If a is a number, define its <u>absolute value</u> $|a|$ by $|a| = \sqrt{a^2}$. Show that $|aA| = |a|\,|A|$, where A is a vector.

2) If A and B are vectors, show that $(|A| + |B|)^2 - (A + B)^2 \geqq 0$. Use this to prove the <u>triangle inequality</u>: $|A + B| \leqq |A| + |B|$. When can this be an equality?

3) In n-space, let $A_1 = (1, 0, \ldots, 0)$, $A_2 = (0, 1, 0, \ldots, 0)$, $\ldots$, $A_n = (0, \ldots, 0, 1)$. Show the following:

a) If $a_1, a_2, \ldots, a_n$ are numbers such that $a_1A_1 + a_2A_2 + \ldots + a_nA_n = 0$, then $a_1 = a_2 = \ldots = a_n = 0$.

b) If B is any vector in n-space, there are numbers $c_1, c_2, \ldots, c_n$ such that $B = c_1A_1 + c_2A_2 + \ldots + c_nA_n$. How many such sets of numbers are there for a given B?

c) $A_i^2 = 1$ and $A_iA_j = 0$ if $i \neq j$, for all i and j.

d) If B is any vector in n-space and $BA_1 = BA_2 = \ldots = BA_n = 0$, then $B = 0$.

e) If A is any vector in n-space, and if $A = a_1A_1 + a_2A_2 + \ldots + a_nA_n$, then $|A|^2 = a_1^2 + a_2^2 + \ldots + a_n^2$ and each $a_i = AA_i$.

4) Show by vectors that the diagonals of a rhombus are perpendicular.

5) Show that the medians of the triangle with vertices P, Q, R meet in $\frac{1}{3}(P + Q + R)$. (Observe that the midpoint of the side $\overline{PQ}$ is $\frac{1}{2}(P + Q)$.)

6) What is the plane through $(-1, 2, -3)$ such that any vector perpendicular to the plane forms equal angles with the vectors $(1, 0, 0)$, $(0, 1, 0)$ and $(0, 0, 1)$?

7) Show that the planes $x - y + 2z = 4$ and $2y - 2x - 4z + 7 = 0$ are parallel. Find the distance between them.

8) Let A and B be non-zero vectors in 3-space. Prove that A and B are parallel if and only if $A \times B = 0$.

9) Prove the law of sines using vectors in 3-space.

10) What is the set of points X in 3-space such that the vector from X to $(1, 3, 5)$ is perpendicular to the vector from $(-5, -1, 3)$ to X? Show that this is the equation of a sphere with center $(-2, 1, 4)$ and radius $\sqrt{14}$.

11) Let B and W be vectors, different from 0. Show that there exist vectors B_1 and B_2 such that

(a) B_1 is parallel to W,
(b) B_2 is perpendicular to W,
(c) $B_1 \cdot B_2 = 0$,

and

(d) $B = B_1 + B_2$.

Show that B_1 and B_2 are unique.

12) Let R and S be the points $(1, 2, 3, 4)$ and $(4, 3, 2, 1)$; let A be the vector $(1, 1, 1, 1)$; let M be the line through R parallel to A; and notice that S does not lie on M.

(a) Given a point Q on M, compute the distance between S and Q and show that it is $\geq 2\sqrt{5}$.

(b) Show that the distance between S and Q is precisely $2\sqrt{5}$ for just one Q_0, being actually $> 2\sqrt{5}$ for all other points Q. Compute the coordinates of this Q_0.

(c) Show that the vector $S - Q_0$ is perpendicular to the vector A.

13) Keeping the results of (2) in mind, consider (in n-space) a point R and a vector A whose length $|A|$ is > 0; let M be the line through R parallel to A; and let S be a point not on M.

(a) Show that there is just one point Q_0 on M such that the vector $S - Q_0$ is perpendicular to A and compute its coordinates.

(b) Given a point Q on M, compute the distance from S to Q and show that it is $\geq |S - Q_0|$. Show that the distance from S to Q is precisely $|S - Q_0|$ if and only if $Q = Q_0$.

14) Compute (in 3-space) the equation of the plane which passes through the points $(1, 0, 0)$, $(0, 1, 0)$, and $(0, 0, 1)$.

(a) Compute the distance $|S - Q|$ between the point $S = (2, 2, 2)$ and a point Q in this plane.

(b) Compute the minimum distance and call it Δ.

(c) Show that there is just one point Q_0 in the plane such that the distance $|S - Q_0| = \Delta$. Compute the coordinates of Q_0.

(d) Show that the vector $S - Q_0$ is perpendicular to the plane.

15) Given three vectors A, B, and C (in n-space) show that the relation

$$(A \cdot B)C = A(B \cdot C)$$

is <u>not</u>, in general, true.

16) A <u>cube</u> in n-dimensional space is a set of points $(x_1, \ldots, x_n)$ satisfying $a_1 \leqq x_1 \leqq a_1 + d$, $a_2 \leqq x_2 \leqq a_2 + d$, $\ldots$, $a_n \leqq x_n \leqq a_n + d$, for some set of n numbers $a_1, \ldots, a_n$ and some $d > 0$. Then the <u>vertices</u> of the cube are the points of the form $(b_1, \ldots, b_n)$, where each b_i is either a_i or $a_i + d$. An <u>edge</u> of the cube is a line segment joining two vertices $(b_1, \ldots, b_n)$ and $(c_1, \ldots, c_n)$, where $b_i = c_i$ for all except one of the i's. How many vertices and edges has a cube in 2-space? in 3-space? in 4-space? in n-space? How would you define a 2-dimensional face of an n-dimensional cube $(n \geqq 2)$? a k-dimensional face $(n \geqq k)$? How many of these are there?

CALCULUS

1. The Derivative

Suppose that we wish to find the <u>tangent</u> to a given curve at a given point of that curve. Before we can attempt to do so, we must know what it <u>means</u> for a line to be tangent to a curve. It certainly means more than that the line intersects the curve only at one point, for this condition is satisfied in the following diagram by the solid line:

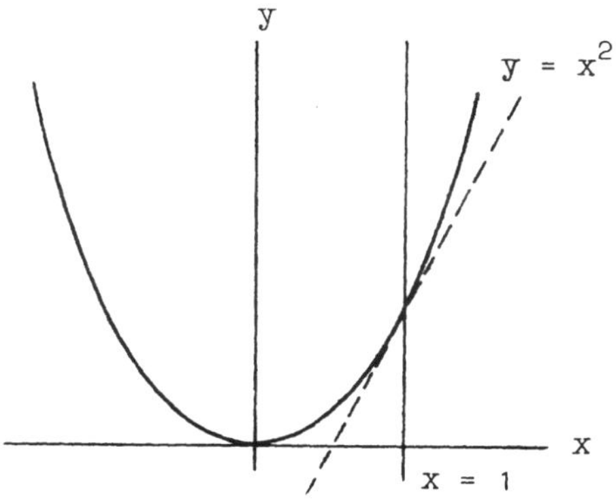

Perhaps one might suggest that the <u>slope</u> of the line is concerned. (If A is a vector parallel to the line, $A = (a_1, a_2)$ the ratio $\frac{a_2}{a_1}$ is the slope of the line if $a_1 \neq 0$; the slope is <u>undefined</u> if $a_1 = 0$. Thus

$$\text{slope} = \frac{\text{vertical rise}}{\text{horizontal run}},$$

if defined.) We could then say the tangent at $x = 1$ to the curve $y = x^2$ is the line through the point $(1, 1)$ and having the same slope as the curve at that point. But now we have substituted the undefined notion of "slope of a curve" for the idea of "tangent to a curve."

Just as we have an intuitive notion which enables us to sketch in the diagram above a (broken) line which approximates the tangent to the curve at a point, we have a notion of slope of the curve at a point. In fact, our intuitive notion of slope is probably just "the slope of the tangent line". We see from this example that the slope at different points, in this sense, may be different; it depends on the point, which in the case of each individual curve, depends on the x-coordinate of the point. Thus the slope is a <u>function</u> of x, i.e., there is a <u>rule</u> (perhaps quite complicated) by which we can assign to each number x of a certain collection (not necessarily <u>all</u> x) a <u>single</u> number $s(x)$, which is in this case the slope of the curve at the given value of x.

But the question, "What is the slope?" remains unanswered, and the reason is a very simple one. For within the language of our algebra and geometry there is no combination of concepts capable of expressing what we intuitively feel that "slope" means. Therefore we must <u>define</u> the slope in terms which involve certain new notions.

Now suppose I want to find the slope of a mountainside at some point. What would I do? I could get an <u>estimate</u> of the slope by standing at the point of interest and sending a helper up the mountain a little way. Then we could compare the difference in height with the horizontal distance between us, giving a ratio which would serve as an approximate slope.

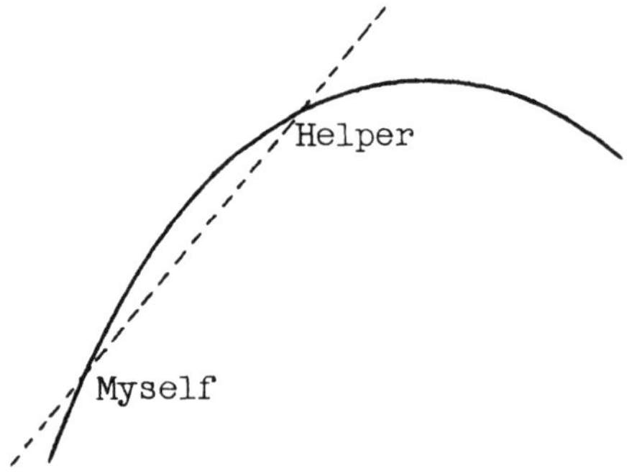

This would be the slope of the broken line in the diagram. It is evident from the diagram that if I were to send my helper very far away, perhaps over the top of the mountain, the approximate values I would get for the slope would probably be meaningless. On the other hand, if he came closer to me, the estimates we would get would be closer to what we would like to call the slope. More estimates could also be obtained by sending him down the mountain, and these would likewise be better if he were not sent so far.

Now let us try this procedure on the graph of the function $y = x^2$ at the point where $x = 1$, i.e., at $(1, 1)$. I stand at $(1, 1)$ and send my helper up to where his x-coordinate is $1 + h$. Then he is at the point $(1 + h, (1 + h)^2)$, because he is on the curve. The approximate value we get for the slope is

$$\frac{(1+h)^2 - 1}{(1+h) - 1} = \frac{2h + h^2}{h} = 2 + h .$$

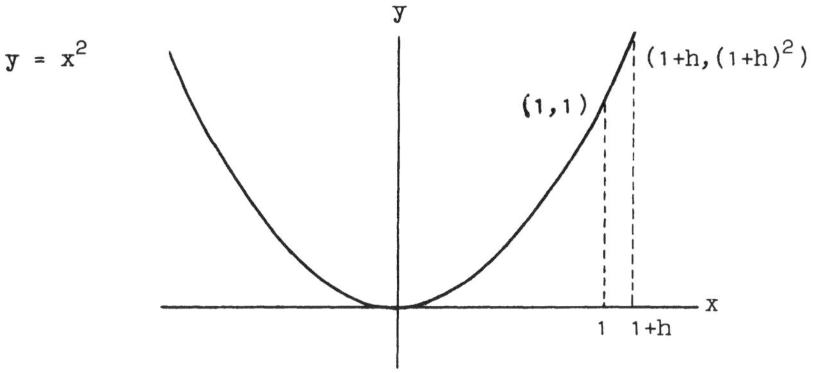

As the distance h that he goes ahead is made shorter, or as h approaches zero, our approximation for the slope becomes closer and closer to 2. This is likewise the case if we send him down the curve, for this simply amounts to using a negative h in the above process. For instance, when $h = .0001$, we get an approximate slope of 2.0001; when $h = -.000001$, an approximate slope of 1.999999. Then we define the slope of $y = x^2$ at $x = 1$ to be 2, which is the number approached by the approximations to the slope as h approaches zero.

We have thus reduced the undefined geometrical term "slope" to the undefined numerical term "approaches." At this stage we shall rest in the process of definition. A precise definition of what is meant by "approaches" will be given later in the course when experience has prepared you for it, but we shall see in examples that it is quite easy to determine the slopes of many common curves without entering deeply into the concepts of "limit" and "approaching." Where special properties of these concepts are used, they should be noted, and careful proofs will be supplied at a later time.

Now let $y = f(x)$ be a function of x, and let x_0 be a certain fixed value for x. Suppose that the approximate slopes, or as we say, the difference quotients,

$$\frac{f(x_0+h) - f(x_0)}{h}$$

approach a single number b as h approaches zero. Then we define the slope of the graph of $y = f(x)$ at x_0, or the derivative of $f(x)$ at x_0, to be the number b. If there is no single number b approached by the difference quotients (for instance if they approach 1 for positive h's approaching zero and -1 for negative h's approaching zero), we say that the slope (or derivative) of $y = f(x)$ at $x = x_0$ is undefined. If it is defined, we denote it by

$$\lim_{h \to 0} \frac{f(x_0+h) - f(x_0)}{h},$$

which we read as "the limit as h goes to zero of

$$\frac{f(x_0+h) - f(x_0)}{h}."$$

For all values of x where $y = f(x)$ has a derivative defined, this derivative is a function of x. It is customarily written in one of several notations, of which we prefer y', or $f'(x)$ ("f-prime of x"). Other notations are $\frac{dy}{dx}$ (not a fraction) and Dy, also $\frac{df}{dx}$ and $Df(x)$. We then have the symbolic definition:

$$y' = f'(x) = \lim_{h \to 0} \frac{f(x+h) - f(x)}{h}$$

if it exists. The process of obtaining $f'(x)$ from $f(x)$ is called differentiation.

2. Examples of Derivatives

a) $y = f(x) = x$. $\frac{f(x+h) - f(x)}{h} = \frac{(x+h) - x}{h} = \frac{h}{h} = 1$.

As $h \to 0$, this is unaffected, hence is regarded as approaching 1 (since it is near 1 when h is near zero). Thus we have

$$y' = f'(x) = 1 \text{ for all } x.$$

b) $y = f(x) = c$, i.e., $f(x)$ is a constant c for all x. (If you prefer to see an "x" in a function of x, you may write $f(x) = (x + c) - x$.)

$$\frac{f(x+h) - f(x)}{h} = \frac{c - c}{h} = 0,$$

which is unchanged as h approaches 0. Thus $f'(x) = 0$ in this case for all x.

c) $y = f(x) = x^n$, n a positive integer.

$$\frac{f(x+h) - f(x)}{h} = \frac{(x+h)^n - x^n}{h}$$

$$= \frac{x^n + nx^{n-1}h + \frac{n(n-1)}{2}x^{n-2}h^2 + \dots + nxh^{n-1} + h^n - x^n}{h},$$

by the binomial theorem. But this is

$$nx^{n-1} + \frac{n(n-1)}{2} x^{n-2}h + \dots + nxh^{n-2} + h^{n-1} .$$

As $h \longrightarrow 0$, all terms but the first go to zero, and their sum likewise goes to zero. (This seems quite obvious, but we shall eventually prove it.) Thus $\underline{f'(x) = nx^{n-1}}$.

d) $y = f(x) = x^{\frac{p}{q}}$ (this is only defined for $x \geqq 0$.)

$$\frac{f(x+h) - f(x)}{h} = \frac{(x+h)^{\frac{p}{q}} - x^{\frac{p}{q}}}{h} .$$

Let $u = (x + h)^{\frac{1}{q}}$, $v = x^{\frac{1}{q}}$. Then as $h \longrightarrow 0$, $u \longrightarrow v$. Also $x + h = u^q$, $x = v^q$, and $h = u^q - v^q$. The difference quotient becomes

$$\frac{u^p - v^p}{u^q - v^q} = \frac{(u-v)(u^{p-1} + u^{p-2}v + \dots + uv^{p-2} + v^{p-1})}{(u-v)(u^{q-1} + u^{q-2}v + \dots + uv^{q-2} + v^{q-1})}$$

$$= \frac{u^{p-1} + u^{p-2}v + \dots + uv^{p-2} + v^{p-1}}{u^{q-1} + u^{q-2}v + \dots + uv^{q-2} + v^{q-1}} .$$

The numerator consists of p terms, each approaching v^{p-1}, and the denominator has q terms, each approaching v^{q-1}. The quotient then approaches

$$\frac{p \cdot v^{p-1}}{q \cdot v^{q-1}} = \frac{p}{q} v^{p-q} = \frac{p}{q} x^{\frac{p-q}{q}} = \frac{p}{q} x^{\frac{p}{q} - 1} .$$

Here we have assumed $x \neq 0$, and that the limit of a quotient is the quotient of the limits if the denominator does not approach zero. Thus

$$f'(x) = \frac{p}{q} x^{\frac{p}{q} - 1} ,$$

and we have proved that if $f(x) = x^a$, a a <u>positive rational number</u>, i.e., a quotient of positive integers, then $\underline{f'(x) = ax^{a-1}}$.

EXERCISES

Write the difference quotients for the following functions $f(x)$.

Find $f'(x)$ when you can.

1) $f(x) = \sin(\log x)$

2) $f(x) = x^{2/3}$

3) $f(x) = |x|$

4) $f(x) = 10^{\tan x} + \frac{x}{Q(x)}$

5) $f(x) = [x]$ ("greatest integer in x", or the largest integer which is $\leqq x$; thus $[\frac{5}{2}] = 2$, $[1] = 1$, $[0] = 0$, $[-\frac{10}{3}] = -4$.)

3. Sums, Products, Reciprocals

Let $f(x)$ and $g(x)$ be functions having derivatives $f'(x)$ and $g'(x)$. Let us look for the derivative of the function $f(x) + g(x)$. The appropriate difference quotient is

$$\frac{(f(x+h) + g(x+h)) - (f(x) + g(x))}{h}$$

$$= \frac{f(x+h) - f(x)}{h} + \frac{g(x+h) - g(x)}{h} \quad .$$

As $h \longrightarrow 0$, the first term goes to $f'(x)$, the second to $g'(x)$, and the sum to $f'(x) + g'(x)$. This gives the rule

$$(*) \qquad \underline{(f(x) + g(x))' = f'(x) + g'(x)} \quad .$$

Now let c be a number, and form the function $cf(x)$. Its difference quotient is

$$\frac{cf(x+h) - cf(x)}{h} = c \cdot \frac{f(x+h) - f(x)}{h} \quad .$$

As $h \longrightarrow 0$, this approaches $cf'(x)$, giving the rule

$$(**) \qquad \underline{(cf(x))' = cf'(x)} \quad .$$

On the basis of the rules (*) and (**) and the examples b) and c), we are ready to calculate the derivative of any <u>polynomial</u>, i.e., any function of the form $a_n x^n + a_{n-1} x^{n-1} + \ldots + a_1 x + a_o$, where $a_o, a_1, \ldots, a_n$ are numbers and n is a positive integer. For by b), the derivative of a_o is 0, and by c) and (**), that of $a_i x^i$ is $i a_i x^{i-1}$ for $1 \leqq i \leqq n$. Thus we can find the derivative of each term in the sum. From (*) we see that the derivative of a sum of two terms is the sum of their derivatives. It follows for three terms, since

$$(f(x) + g(x) + h(x))' = ((f(x) + g(x)) + h(x))'$$

$$= (f(x) + g(x))' + h'(x) = (f'(x) + g'(x)) + h'(x)$$

$$= f'(x) + g'(x) + h'(x).$$

Note that we have simply used (*) twice. A repetition shows that the rule holds for four terms, then for five, and so on for any desired number. This is an occasion for the use of the principle of induction, about which we shall say more in the next section. We have therefore shown that if

$$f(x) = a_n x^n + a_{n-1}x^{n-1} + \dots + a_1 x + a_0 ,$$

then

$$f'(x) = na_n x^{n-1} + (n-1)a_{n-1}x^{n-2} + \dots + 2a_2 x + a_1 .$$

Thus we can differentiate any polynomial.

NOTE. The procedure by which the rule for polynomials was derived deserves some analysis. Knowing only the derivatives of the powers x^k of x, k an integer $\geqq 0$, we were able to determine the derivative of any combination of these with constant coefficients. This was really a result of the rules (*) and (**). The property expressed by these rules is called the linearity of differentiation. In general, an operation performed on functions is called linear if the result of operating on the sum of two functions with it is the same as forming the sum of the results of operating on the two functions separately, and if the result of operating on a constant multiple of a function is the same constant times the result of operating on the function. This property is very important in mathematics, and we shall see more occurrences of it as we proceed.

Now consider the function $f(x)g(x)$. Its difference quotient is

$$\frac{f(x+h)g(x+h) - f(x)g(x)}{h}$$

and there is no obvious direct way of attacking this. We must therefore resort to somewhat indirect means, introducing a trick which is often useful. If the difference quotient were

$$\frac{f(x+h)g(x+h) - f(x+h)g(x)}{h} = f(x + h) \cdot \frac{g(x+h) - g(x)}{h} ,$$

this would at least contain an expression whose limit we recognize. But if we had used this in place of the original difference quotient, we should have had to correct it by adding their difference, or

$$\frac{f(x+h)g(x) - f(x)g(x)}{h} = \frac{f(x+h) - f(x)}{h} \cdot g(x) .$$

Now we have to consider the limit as $h \longrightarrow 0$ of

$$\frac{f(x+h)g(x+h) - f(x)g(x)}{h} = \underset{\substack{\downarrow \\ f(x)}}{f(x + h)} \cdot \underset{\substack{\downarrow \\ g'(x)}}{\frac{g(x+h) - g(x)}{h}} + \underset{\substack{\downarrow \\ f'(x)}}{\frac{f(x+h) - f(x)}{h}} \cdot \underset{\substack{\downarrow \\ g(x)}}{g(x)} .$$

But as $h \longrightarrow 0$, the quantities involved approach the expressions indicated by arrows below them. The only questionable matter here may be $f(x + h) \longrightarrow f(x)$; this expresses a property called <u>continuity</u> which is possessed by all functions which have derivatives, and by many others as well. This will be discussed more fully in the next paragraph. Thus the whole expression approaches $f(x)g'(x) + f'(x)g(x)$, and we have the rule

(***) $$\underline{(f(x)g(x))' = f(x)g'(x) + g(x)f'(x)} .$$

A function $f(x)$ is called <u>continuous at</u> x_o if whenever x is near x_o, then $f(x)$ is near $f(x_o)$, or in other words, if

$$\lim_{h \longrightarrow 0} f(x_o + h) = f(x_o).$$

Now suppose $f'(x_o)$ exists. Then

$$\frac{f(x_o+h) - f(x_o)}{h} - f'(x_o) = \epsilon(h) ,$$

where $\epsilon(h) \longrightarrow 0$ as $h \longrightarrow 0$, and

$$f(x_o + h) - f(x_o) = \underset{\substack{\downarrow\downarrow \\ 0f'(x_o)}}{hf'(x_o)} + \underset{\substack{\downarrow \\ 0}}{h}\, \underset{\substack{\downarrow \\ 0}}{\epsilon(h)} .$$

As $h \longrightarrow 0$, the arrows indicate the limits on the right-hand side. Thus

$$\lim_{h \to 0} (f(x_0 + h) - f(x_0)) = 0,$$

or

$$\lim_{h \to 0} f(x_0 + h) = f(x_0).$$

Therefore _if_ $f'(x_0)$ _exists, then_ $f(x)$ _is continuous at_ x_0 .

Next consider the function $\frac{1}{f(x)}$, where $f(x)$ is never zero. The difference quotient is

$$\frac{\frac{1}{f(x+h)} - \frac{1}{f(x)}}{h} = \frac{\frac{f(x) - f(x+h)}{f(x+h) \cdot f(x)}}{h}$$

$$= - \underset{\displaystyle f'(x)}{\underset{\downarrow}{\frac{f(x+h) - f(x)}{h}}} \cdot \underset{\displaystyle \frac{1}{f(x)}}{\underset{\downarrow}{\frac{1}{f(x+h)}}} \cdot \underset{\displaystyle \frac{1}{f(x)}}{\underset{\downarrow}{\frac{1}{f(x)}}} .$$

As $h \to 0$, this approaches

$$- \frac{f'(x)}{(f(x))^2} ,$$

in the manner indicated by the arrows. We have therefore derived

$$(****) \qquad \underline{\left(\frac{1}{f(x)}\right)' = \frac{-f'(x)}{(f(x))^2}} .$$

It follows that

$$\left(\frac{f(x)}{g(x)}\right)' = \left(f(x) \cdot \left(\frac{1}{g(x)}\right)\right)' = f(x) \cdot \frac{-g'(x)}{(g(x))^2} + f'(x) \cdot \frac{1}{g(x)} ,$$

or

$$\underline{\left(\frac{f(x)}{g(x)}\right)' = \frac{f'(x)g(x) - f(x)g'(x)}{(g(x))^2}} .$$

4. Mathematical Induction

The principle of induction which we mentioned before is a statement about the positive integers. It reads as follows:

Principle of Induction. Let $P(n)$ be an assertion about the positive integer n. Suppose that:

(a) The assertion $P(1)$ is true;

(b) If the assertion $P(k)$ is true for a given integer k, then the assertion $P(k+1)$ is also true.

Then the assertion $P(n)$ is true for all n.

If we think about this a little, we should see that it is entirely in keeping with our intuition. There are various models to help us visualize the principle. In one such model, we consider a system of dominoes on edge, aligned so that if any one tips toward the next one it will knock it over toward the following one. This is the analogue of the hypothesis (b). The analogue of (a) is simply that we can knock over the first one. But then it will tip over the second, which will hit the third, ... By writing enough such steps, we could reach any specified domino, or the assertion $P(n)$ for any prescribed n. However, we have now seen that the hypotheses (a) and (b) are the essentials of this process, and that we can save all this writing simply by showing that they are valid.

As an example, let us give a proof by induction of a formula for the sum of a familiar arithmetic progression:

$$1 + 2 + \dots + n = \frac{n(n+1)}{2} .$$

Then $P(n)$ is the statement that the sum of the first n integers is equal to $\frac{n(n+1)}{2}$. $P(1)$ thus says that the sum of the first integer, or 1, is equal to $\frac{1(1+1)}{2} = 1$. Therefore hypothesis (a) is satisfied. Now hypothesis (b) begins with, "If $P(k)$ is true,". This means that we must suppose $P(k)$ to be true, and see if the truth of $P(k+1)$ is a consequence of this assumption. Thus we begin by supposing

$$P(k) : 1 + 2 + \dots + k = \frac{k(k+1)}{2} .$$

Then we must consider the sum

$$1 + 2 + \dots + k + (k+1) = (1 + 2 + \dots + k) + (k+1) .$$

But, by what we are assuming,

$$1 + 2 + \dots + k = \frac{k(k+1)}{2} = (k+1) \cdot \frac{k}{2} .$$

Therefore

$$1 + 2 + \dots + k + (k+1) = (k+1)\left(\frac{k}{2} + 1\right) = (k+1)\left(\frac{k+2}{2}\right) ,$$

or

$$1 + 2 + \dots + (k+1) = \frac{(k+1)((k+1) + 1)}{2} ,$$

which is $P(k + 1)$. Thus (a) and (b) are satisfied, and $P(n)$ holds for all n by induction.

EXERCISES

1) Prove that $1^2 + 2^2 + \dots + n^2 = \frac{n(n+1)(2n+1)}{6}$.
2) Prove that $(f_1 + f_2 + \dots + f_n)' = f_1' + f_2' + \dots + f_n'$.
3) Prove that $(f_1 f_2 \dots f_n)' = f_1' f_2 \dots f_n + f_1 f_2' f_3 \dots f_n + \dots + f_1 \dots f_{n-1} f_n'$.
4) Show that hypothesis (b) is satisfied for $P(n) : n = n + 2$, which is obviously false.
5) Give an example of a $P(n)$ which is false for $n > 1$ but for which hypothesis (a) holds.
6) Prove that $1 + 3 + 5 + \dots + (2n-1) = n^2$, $n = 1, 2, 3, \dots$.
7) Prove that $(1 - \frac{1}{4})(1 - \frac{1}{9})(1 - \frac{1}{16}) \dots (1-n^{-2}) = \frac{n+1}{2n}$ $n = 2, 3, 4, \dots$.

5. The Chain Rule

Let $f(x)$ and $g(x)$ have derivatives, and consider the composite function $f(g(x))$. We shall develop a rule for use in differentiating this function, and this rule will greatly extend the collection of functions which we can differentiate. For example, the function $y = \sqrt{x^2 + 1}$ can be written as $f(g(x))$, where $f(u) = \sqrt{u}$, $g(x) = x^2 + 1$, and we can differentiate each of these functions.

The difference quotient we must consider is

$$\frac{f(g(x+h)) - f(g(x))}{h} .$$

If we consider the range of $g(x)$, that is, the set of values of $g(x)$, as a set of numbers of which we denote the general one by u, we first consider the difference

$$\frac{f(u+k) - f(u)}{k} - f'(u) \quad .$$

For k small, this expression is close to 0. We put it over a common denominator and write

$$\epsilon(k) = \frac{f(u+k) - f(u) - kf'(u)}{k} \quad .$$

Then $\epsilon(k)$ is a number depending on k, and is small when k is small. Of course, $k = 0$ is forbidden. Let us rewrite this equation as

$$f(u + k) - f(u) = k\epsilon(k) + kf'(u) \; .$$

If we were to choose any number at all and call it $\epsilon(0)$, this last equation would be true for all k, including $k = 0$. But in order to make $\epsilon(k)$ small with k, we decide to choose the number 0 as the one which we shall call $\epsilon(0)$. Thus we define $\epsilon(0) = 0$.

In the original difference quotient, let us write $u = g(x)$, $u + k = g(x + h)$. Then $k = g(x + h) - g(x)$. By the continuity of $g(x)$, $k \longrightarrow 0$ as $h \longrightarrow 0$, and so $\epsilon(k) \longrightarrow 0$ as $h \longrightarrow 0$. Now we have

$$\frac{f(g(x+h)) - f(g(x))}{h} = \frac{f(u+k) - f(u)}{h} = \frac{k\epsilon(k) + kf'(u)}{h}$$

$$= \underset{\displaystyle g'(x)}{\underset{\displaystyle\downarrow}{\frac{g(x+h) - g(x)}{h}}} \cdot \underset{\displaystyle 0}{\underset{\displaystyle\downarrow}{\epsilon(k)}} + \underset{\displaystyle g'(x)}{\underset{\displaystyle\downarrow}{\frac{g(x+h) - g(x)}{h}}} \cdot \underset{\displaystyle f'(u)}{\underset{\displaystyle\downarrow}{f'(u)}} \quad .$$

The quantities involved approach the limits indicated as $h \longrightarrow 0$. Therefore the whole expression approaches $f'(u)g'(x)$, and we have the chain rule:

$$\underline{(f(g(x)))' = f'(g(x)) \cdot g'(x)} \quad .$$

EXERCISE

Develop a rule for $(f_1(f_2(\ldots (f_n(x)) \ldots)))'$, and prove it by induction.

6. Other Functions and Their Derivatives

For the sake of developing technique, we mention several other functions and give their derivatives. Satisfactory definitions of the sine and cosine from the mathematical point of view will not be given in this course, but we shall present intuitive evidence for the formulas for their derivatives. The exponential and logarithmic functions will be discussed in more detail at a later stage in the course.

Function	Derivative
$\sin x$ (x measured in radians)	$\cos x$
$\cos x$	$-\sin x$
e^x	e^x
$\log x$ ($= \log_e x$)	$\frac{1}{x}$

Here e is the number 2.71828..., the base of the system of so-called natural logarithms. Whenever we write $\log x$, it will mean the logarithm to the base e. Now we can derive a rule for differentiating a^x, where a is any fixed positive number. But first we must define a^x for arbitrary exponent x. Assuming that the exponential and logarithm functions are defined, we define $a^x = e^{x \log a}$. Then if we set $f(u) = e^u$ ($f'(u) = e^u$ also), $g(x) = x \log a$, we have $a^x = f(g(x))$, and

$$(a^x)' = f'(g(x)) \cdot g'(x) = e^{x \log a} \cdot \log a = a^x \log a .$$

NOTE. The functions x^b, where b is fixed, and a^x, where $a > 0$ is fixed, are of a completely different nature, and one should bear this in mind when forming the derivatives of expressions involving them. The derivative of x^b is bx^{b-1}; that of a^x is $a^x \log a$.

Let us now form the difference quotients for $\sin x$ and $\cos x$. They are:

$$\frac{\sin(x+h) - \sin x}{h} = \frac{\sin x \cos h + \cos x \sin h - \sin x}{h}$$

and

$$\frac{\cos(x+h) - \cos x}{h} = \frac{\cos x \cos h - \sin x \sin h - \cos x}{h} ,$$

or

(1) $$\frac{\cos h - 1}{h} \cdot \sin x + \frac{\sin h}{h} \cdot \cos x ,$$

and

(2) $$\frac{\cos h - 1}{h} \cdot \cos x - \frac{\sin h}{h} \cdot \sin x .$$

Thus we are led to consider

$$\lim_{h \to 0} \frac{\cos h - 1}{h} \quad \text{and} \quad \lim_{h \to 0} \frac{\sin h}{h} .$$

We can reduce the first of these to the second as follows:

$$\frac{\cos h - 1}{h} = \frac{\cos^2 h - 1}{h(\cos h + 1)} = \frac{-\sin^2 h}{h(\cos h + 1)} = \frac{\sin h}{h} \cdot \sin h \cdot \frac{-1}{\cos h + 1} .$$

Of these three factors, the second goes to 0 as $h \to 0$, and the third approaches $-\frac{1}{2}$. If the limit of $\frac{\sin h}{h}$ exists at all, the limit of $\frac{\cos h - 1}{h}$ will therefore be 0.

Now consider the diagram on the right. The area of the triangle OAB is less than or equal to that of the sector OCB, which in turn is less than or equal to that of triangle OCD. But we know:

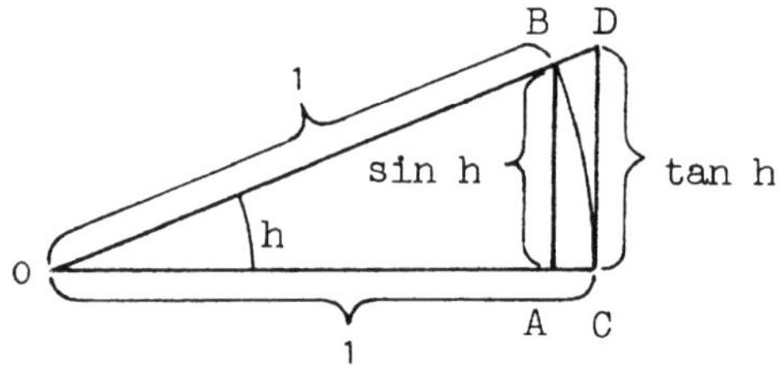

$$OAB = \frac{1}{2} \sin h \cdot \cos h$$

sector $OCB = \frac{h}{2}$ (since the area of the whole circle of radius 1 is π, and there are 2π radians in the circle.)

$$OCD = \frac{1}{2} \tan h = \frac{1}{2} \frac{\sin h}{\cos h} .$$

Thus

$$\frac{1}{2} \sin h \cdot \cos h \leqq \frac{h}{2} \leqq \frac{1}{2} \frac{\sin h}{\cos h} .$$

If $h > 0$, the first inequality gives

$$\frac{\sin h}{h} \leqq \frac{1}{\cos h} ,$$

and the second gives

$$\cos h \leqq \frac{\sin h}{h} .$$

Thus

$$\cos h \leqq \frac{\sin h}{h} \leqq \frac{1}{\cos h} .$$

But as $h \longrightarrow 0$, $\cos h$ and $\frac{1}{\cos h}$ both approach 1. Since $\frac{\sin h}{h}$ lies between these quantities, it must also approach 1. A similar argument can be used for $h < 0$. Therefore we conclude that

$$\lim_{h \longrightarrow 0} \frac{\sin h}{h} = 1, \qquad \lim_{h \longrightarrow 0} \frac{\cos h - 1}{h} = 0 .$$

Taking limits in the difference quotients (1) and (2), we obtain the results originally given for the derivatives of the sine and cosine.

(NOTE. Owing to the presence of undefined terms, you should regard this argument not as a proof, but rather as an effort to convince you that our statements are reasonable in the light of your experience.)

EXERCISES

A number of exercises involving use of the rules for differentiation so far developed are to be found in any standard calculus text. We list, as a suggestion, a few problems. In many cases, these will not provide sufficient practice in technique; then further practice is of course required.

1. Differentiate: $\tan x$, $\cot x$, $\sec x$, $\csc x$.
2. R. Courant, Differential and Integral Calculus, v. I: p. 109, ex. 2; p. 144, ex. 5, 6, 7, 8, 9; p. 157, ex. 1-15; p. 177, ex. 2, 3, 4, 5, 9, 12, 13.
3. C. O. Oakley, The Calculus; Barnes and Noble College Outlines: p. 43, ex. 1-13, 16, 17, 19, 20.

7. Rolle's Theorem, Mean Value Theorem, Maxima and Minima

We have mentioned continuous functions before, but let us restate the preliminary definition which we have used. Let $f(x)$ be a function which is defined in an interval containing x_0. Then we say that $f(x)$ is continuous at x_0 if changing x only slightly from x_0 produces only a small change in the value of $f(x)$ from its value $f(x_0)$ at x_0. More briefly, if x is near x_0, then $f(x)$ is near $f(x_0)$. Please observe that this definition is imprecise in that the words <u>small</u> and <u>near</u> have no exact meaning. For one man, within .01 might be near enough, while another might consider a number as small until it is less than .00001. The precise definition of continuity which we eventually shall give will satisfy both men and all others as well.

Now let $f(x)$ be defined for all x such that $a \leqq x \leqq b$. A point x_0 of this interval is called a <u>maximum of</u> $f(x)$ <u>in the interval</u> if $f(x_0) \geqq f(x)$ for all x such that $a \leqq x \leqq b$. x_1 is called a <u>minimum of</u> $f(x)$ <u>in the interval</u> if $f(x_1) \leqq f(x)$ for $a \leqq x \leqq b$. Not all functions need have maxima or minima; for consider the function

$$\left\{\begin{array}{ll} f(x) = x, & 0 \leqq x < 1 \\ f(1) = 0 & \end{array}\right\}$$

Its graph is

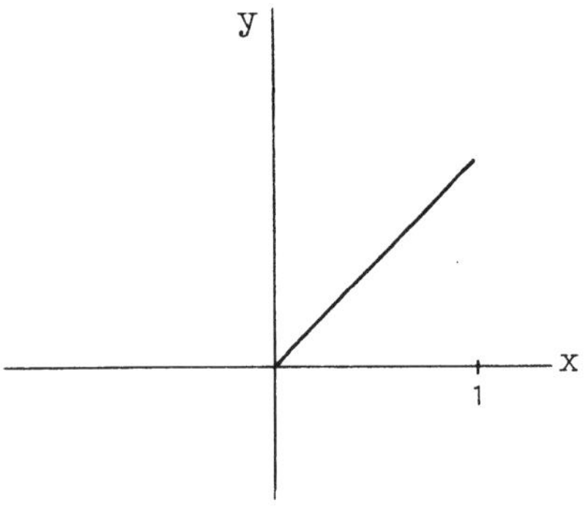

where the oblique line has no right-hand endpoint. This function has <u>no maximum</u> in the interval $0 \leqq x \leqq 1$. For surely 1 cannot be a maximum since $f(1) = 0 < f(\frac{1}{2})$. But if $a < 1$ were to be a maximum observe that

$$a < \frac{a+1}{2} < 1 .$$

Therefore

$$a = f(a) < f\left(\frac{a+1}{2}\right) = \frac{a+1}{2} ,$$

so that a cannot be a maximum. Note, however, that this function is <u>not continuous</u> at 1.

Later in the course, we shall prove the following theorem:

THEOREM. Let $f(x)$ be continuous at every point of the interval $a \leqq x \leqq b$. Then there is at least one x_0, $a \leqq x_0 \leqq b$, such that x_0 is a maximum for $f(x)$ in $a \leqq x \leqq b$.

Observe that it follows that $f(x)$ has a <u>minimum</u> in $a \leqq x \leqq b$. For a minimum of $f(x)$ is simply a maximum of $-f(x)$.

Now let $f(x)$ have a maximum at x_0, $a < x_0 < b$, so that x_0

is not an endpoint. Then consider the difference quotient

$$\frac{f(x_0+h) - f(x_0)}{h} .$$

For all h,

$$f(x_0 + h) - f(x_0) \leqq 0.$$

If $h > 0$,

$$\frac{f(x_0+h) - f(x_0)}{h} \leqq 0 .$$

If $h < 0$,

$$\frac{f(x_0+h) - f(x_0)}{h} \geqq 0 .$$

If $f'(x_0)$ exists, these expressions approach the common limit $f'(x_0)$ as $h \longrightarrow 0$. But if $f'(x_0) > 0$, the first one (for $h > 0$) could never get closer to $f'(x_0)$ than the distance of $f'(x_0)$ from 0. If $f'(x_0) < 0$, the second expression $(h < 0)$ could not approach it. Therefore we must have $f'(x_0) = 0$. Our conclusion is then:

Let $f(x)$ be defined for $a \leqq x \leqq b$ and let x_0 be a point such that $a < x_0 < b$ and such that x_0 is a maximum for $f(x)$ in $a \leqq x \leqq b$. Then if $f'(x_0)$ exists, we must have $f'(x_0) = 0$. (We could equally well have used "minimum" instead of "maximum".)

ROLLE'S THEOREM. Let $f(x)$ be continuous in the interval $a \leqq x \leqq b$. Let $f(x)$ be differentiable for all x such that $a < x < b$. If $f(a) = f(b)$, then there is a ξ, $a < \xi < b$, such that $f'(\xi) = 0$.

For smooth-looking functions this seems evident, but we must still prove it. It follows quite quickly from the theorem on the existence of a maximum and our remarks on the derivative at a maximum.

Proof of Rolle's Theorem. Since $f(x)$ is continuous, $f(x)$ has a maximum and a minimum in $a \leqq x \leqq b$. Now either $f(x)$ is constant for $a \leqq x \leqq b$, or $f(x) > f(a) = f(b)$ for some x, or $f(x) < f(a) = f(b)$ for some x. In the first case, we have seen that $f'(x) = 0$ whenever $a \leqq x \leqq b$, and we can take ξ to be any point such that $a < \xi < b$. In the second case, any maximum x_0 of $f(x)$ has $a < x_0 < b$, so $f'(x_0)$ exists. But we have seen that then $f'(x_0) = 0$, and we can take $\xi = x_0$. If the second case does not hold, and $f(x)$ is not constant, then any minimum x_1 of $f(x)$ has $a < x_1 < b$, and $f'(x_1) = 0$ as before. In this case we can take $\xi = x_1$. Thus the theorem is proved.

Next we prove a consequence of Rolle's Theorem, called the Mean Value Theorem, which is perhaps the most useful result we have had to date.

MEAN VALUE THEOREM. Let $f(x)$ be continuous for $a \leqq x \leqq b$, differentiable for $a < x < b$. Then there exists a ξ, $a < \xi < b$, such that

$$f(b) - f(a) = f'(\xi)(b - a) .$$

PROOF. We wish to form a new function $g(x)$ which will satisfy the conditions of Rolle's Theorem and such that the conclusion of Rolle's Theorem for $g(x)$ will imply the desired result. Therefore we try

$$g(x) = f(x)(b - a) - x(f(b) - f(a)) .$$

Then $g(a) = g(b) = f(a) \cdot b - f(b) \cdot a$, $g(x)$ is continuous for $a \leqq x \leqq b$, because $f(x)$ is, and

$$g'(x) = f'(x)(b - a) - (f(b) - f(a)), \quad a < x < b,$$

so that $g(x)$ is differentiable for $a < x < b$. By Rolle's Theorem, there is a ξ, $a < \xi < b$, such that

$$g'(\xi) = 0 = f'(\xi)(b - a) - (f(b) - f(a)),$$

or

$$f(b) - f(a) = f'(\xi)(b - a),$$

as desired. Thus the Mean Value Theorem is proved.

Geometrically, the mean value theorem means that if we form a secant to the graph of a function satisfying the conditions, there is a point between the intersections of the secant with the graph where the tangent to the curve is parallel to this secant.

We shall say considerably more about the uses of the Mean Value Theorem as we go along: we shall likewise discuss questions of maxima and minima at greater length a little later. Let us outline at the moment the procedures we can use for determining maxima and minima of a function $f(x)$ which is continuous for $a \leqq x \leqq b$, and is differentiable for $a < x < b$. We know that if $f(x)$ has a maximum (or minimum) at x_0,

Geometrical meaning of the Mean Value Theorem

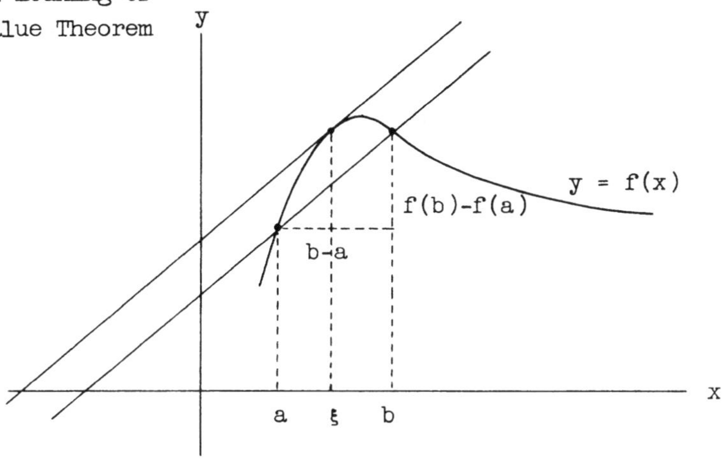

with $a < x_o < b$, then $f'(x_o) = 0$. It is also conceivable that a maximum, minimum, or both might occur at the endpoints a and b. Therefore let $x_1, x_2, \dots, x_n$ be all points within the interval where $f'(x) = 0$, and consider the numbers

$$f(a),\ f(b),\ f(x_1),\ f(x_2),\ \dots,\ f(x_n)\ .$$

Any value of x corresponding to the largest of these is a maximum; a value corresponding to the smallest is a minimum. It may be, therefore, that there are points x_o with $f'(x_o) = 0$ which are neither maxima nor minima for $f(x)$.

EXERCISES

Find a maximum and a minimum for the following functions on the corresponding intervals:

a) $f(x) = x^3 - 6x, \quad 0 \leqq x \leqq 7$.

b) $f(x) = x, \quad -7 \leqq x \leqq 17$.

c) $f(x) = \begin{Bmatrix} x^2 \sin \frac{1}{x}, & x \neq 0 \\ 0 \text{ if } & x = 0 \end{Bmatrix}, \ -\frac{2}{\pi} \leqq x \leqq \frac{2}{\pi}$.

8. Applications of the Mean Value Theorem

We say a function $f(x)$ is strictly increasing if whenever $a < b$, $f(a) < f(b)$. $f(x)$ is strictly decreasing if $a < b$ implies $f(a) > f(b)$. $f(x)$ is called monotone increasing if $a < b$ implies $f(a) \leqq f(b)$, and monotone decreasing if $a < b$ implies $f(a) \geqq f(b)$.

Now suppose $f(x)$ is continuous for $a \leqq x \leqq b$ and $f'(x)$ exists and is positive for all x such that $a < x < b$. Then let c and d be such that $a \leqq c < d \leqq b$. Then surely $f(x)$ satisfies the hypotheses of the Mean Value Theorem on the interval $c \leqq x \leqq d$. Therefore

$$f(d) - f(c) = f'(\xi)(d - c), \quad \text{where} \quad c < \xi < d.$$

But then $f'(\xi) > 0$, and $d - c > 0$. Therefore $f(d) - f(c) > 0$, or $f(c) < f(d)$. This is exactly the condition that $f(x)$ be strictly increasing on $a \leqq x \leqq b$.

In similar fashion, one shows that

a) If $f'(x) < 0$ for $a < x < b$, then $f(x)$ is strictly decreasing.
b) If $f'(x) \geqq 0$ for $a < x < b$, then $f(x)$ is monotone increasing.
c) If $f'(x) \leqq 0$ for $a < x < b$, then $f(x)$ is monotone decreasing.

These are left as exercises. Observe, however, that there are functions which are, for instance, monotone decreasing, but which are not continuous, therefore do not even have derivatives throughout an interval. As an example, let

$$f(x) = \left\{ \begin{array}{ll} 1 - x, & 0 \leqq x < \frac{1}{2} \\ 0\ , & \frac{1}{2} \leqq x \leqq 1 \end{array} \right\} .$$

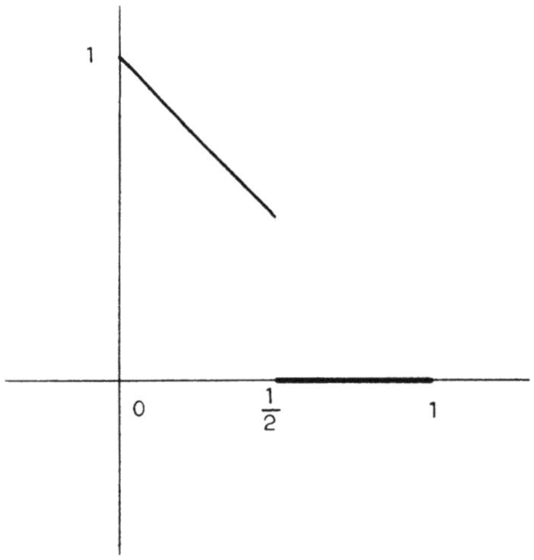

We have seen that the derivative of a constant is zero. Now we can prove the converse, namely that a continuous function $f(x)$ on $a \leqq x \leqq b$ whose derivative $f'(x) = 0$ for all x, $a < x < b$, is a constant. For if $a < d \leqq b$, then $f(d) - f(a) = f'(\xi)(d - a)$ for some ξ, $a < \xi < d$, by the Mean Value Theorem. But then $f'(\xi) = 0$, and $f(d) = f(a)$. Since d could be any number of the interval, we have $f(x) = f(a)$ for all x, $a \leqq x \leqq b$, and $f(x)$ is a constant.

As a corollary to this result,

let $f(x)$ and $g(x)$ be continuous on $a \leqq x \leqq b$ and let $f'(x) = g'(x)$ for all x, $a < x < b$. Then $f(x) - g(x)$ is continuous and has derivative 0, therefore is a constant c. So $f(x) = g(x) + c$. We have proved that two functions having equal derivatives in an interval differ by a constant function. (Note that it was the linearity of the operation of differentiation that made this an easy consequence of the preceding result.)

Next consider the function $f(x) = x - \sin x$. This is continuous and differentiable in any interval; its derivative is $1 - \cos x \geqq 0$. Therefore it is monotone increasing. But $f(0) = 0$, so if $x \geqq 0$, then $f(x) \geqq f(0) = 0$, and $x - \sin x \geqq 0$ if $x \geqq 0$. Likewise the function $-1 + \frac{1}{2}x^2 + \cos x$ has value 0 when $x = 0$; its derivative is $x - \sin x \geqq 0$ for $x \geqq 0$ by the above. Hence it is also a monotone increasing function, and so $-1 + \frac{1}{2}x^2 + \cos x \geqq 0$ when $x \geqq 0$. Now

$$-x + \frac{x^3}{6} + \sin x \geqq 0$$

for $x \geqq 0$, by similar reasoning; continuing, we have

$$1 - \frac{x^2}{2!} + \frac{x^4}{4!} - \cos x \geqq 0,$$

$$x - \frac{x^3}{3!} + \frac{x^5}{5!} - \sin x \geqq 0,$$

$$-1 = \frac{x^2}{2!} - \frac{x^4}{4!} + \frac{x^6}{6!} + \cos x \geqq 0,$$

$$\vdots$$

for all $x \geqq 0$, where $n!$ (n-factorial) is the product $1 \cdot 2 \cdot \ldots \cdot n$. (Note that $2! = 2$, $1! = 1$. We agree to write $0! = 1$.) These can be transposed to give the following inequalities for all $x \geqq 0$:

$$\sin x \leqq x$$

$$1 - \frac{1}{2!}x^2 \leqq \cos x \leqq 1$$

$$\frac{x}{1!} - \frac{x^3}{3!} \leqq \sin x \leqq x$$

$$1 - \frac{x^2}{2!} \leqq \cos x \leqq 1 - \frac{x^2}{2!} + \frac{x^4}{4!}$$

$$\frac{x}{1!} - \frac{x^3}{3!} \leqq \sin x \leqq \frac{x}{1!} - \frac{x^3}{3!} + \frac{x^5}{5!}$$

Thus we have polynomials for approximating $\sin x$ and $\cos x$, at least for $x \geqq 0$ (one can also apply the method to the case $x \leqq 0$). The error in using one of these polynomials as an approximation is at most the difference between it and the next polynomial in the sequence; for example, if we use $1 - \frac{1}{3!} = .833 \ldots$ to estimate $\sin 1$, our error is at most $\frac{1}{5!} = \frac{1}{120}$.

PROBLEMS

1. Let $f(x) = \cos 2x$, $g(x) = 2 \cos^2 x$. Find $f'(x)$ and $g'(x)$ and show that $f'(x) = g'(x)$. Then find a constant c such that $f(x) = g(x) + c$.
2. Courant, p. 557, ex. 77.
3. Estimate $\sin (\frac{1}{2})$ to five decimal places.

9. Curve Tracing. Tangent and Normal Lines

Now we shall see how some of the things we have learned can help us to get a quick idea of the appearance of the graph of a function $f(x)$. In practice, we are usually concerned with functions which have as many derivatives as we wish (the second derivative $f''(x)$ is simply the derivative of the derivative $f'(x)$, and so on to higher derivatives) except at a few points whose exceptional behavior is fairly clear from the start.

To begin with, we must extend our definition of maximum and minimum to include relative maxima and minima. Let $f(x)$ be defined in the interval $a \leqq x \leqq b$. A point x_0, $a < x_0 < b$, is called a relative maximum of $f(x)$ if there is a neighborhood of x_0, that is, an interval $c \leqq x \leqq d$ such that $a \leqq c < d \leqq b$ and $c < x_0 < d$, with the property that x_0 is a maximum for $f(x)$ in this interval. Relative minimum is similarly defined. If $f(x)$ is continuous in $a \leqq x \leqq b$ and differentiable for $a < x < b$, the same properties hold in any subinterval $c \leqq x \leqq d$; our earlier considerations of maxima and minima then tell us that $f'(x_0) = 0$ whenever x_0 is a relative maximum or a relative minimum.

The converse is false, as is easily seen by considering the curve $y = x^3$ at $x_0 = 0$. Here $y' = 3x^2$, so $y'(0) = 0$, but 0 is neither a relative maximum nor a relative minimum, as is evident from the graph. Therefore once we have determined the points where $y' = 0$, we must determine whether they are relative maxima, relative minima, or neither of these.

Suppose $f'(x_0) = 0$, and suppose we also calculate $f''(x_0)$ and find $f''(x_0) > 0$. Then for small h,

$$\frac{f'(x_0+h) - f'(x_0)}{h} = \frac{f'(x_0+h)}{h} > 0.$$

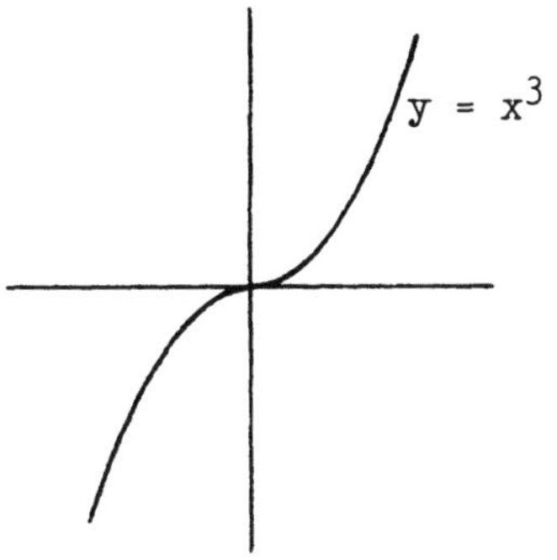

Thus if $h > 0$, $f'(x_0 + h) > 0$; if $h < 0$, $f'(x_0 + h) < 0$. Now consider an interval $x_0 - \delta \leqq x \leqq x_0 + \delta$ small enough so that for all h such that $0 < h \leqq \delta$, $f'(x_0 + h) > 0$ and for all h such that $-\delta \leqq h < 0$, $f'(x_0 + h) < 0$. Then x_0 <u>is a minimum for</u> $f(x)$ <u>in this interval</u>. For if x_1 is any point of the interval and $x_1 > x_0$, then $f(x_1) - f(x_0) = f'(\xi)(x_1 - x_0)$, where $x_1 > \xi > x_0$, or $\xi = x_0 + (\xi - x_0)$, with $0 < \xi - x_0 < \delta$. Thus $f'(\xi) > 0$, and $f(x_1) > f(x_0)$. In similar fashion we show that if x_2 is a point of the smaller interval and $x_2 < x_0$, then $f(x_2) > f(x_0)$. This proves the assertion, and shows in addition that x_0 is the <u>only</u> minimum of $f(x)$ in $x_0 - \delta \leqq x \leqq x_0 + \delta$. We have now proved the following:

<u>If</u> $f'(x_0) = 0$ <u>and if</u> $f''(x_0) > 0$, <u>then</u> x_0 <u>is a relative minimum for</u> $f(x)$.

By the same type of reasoning, we can prove:

<u>If</u> $f'(x_0) = 0$ <u>and if</u> $f''(x_0) < 0$, <u>then</u> x_0 <u>is a relative maximum for</u> $f(x)$.

If $f'(x_0) = 0 = f''(x_0)$, our information is still incomplete; we could have behavior like that of $y = x^3$ at 0, or a minimum as in the case of $y = x^4$ at 0, or a maximum as in the case of $y = -x^4$ at 0. We could even have more complicated behavior; consider the following function at $x = 0$:

$$f(x) = \left\{ \begin{array}{ll} x^4 \sin \frac{1}{x}, & x \neq 0 \\ 0 \quad , & x = 0 \end{array} \right\} .$$

Here $f'(0) = 0 = f''(0)$, but the curve crosses its tangent (the x-axis) infinitely many times in every neighborhood of 0. This curve, however, is really impossible to sketch with any accuracy in a neighborhood of 0 because the breadth of the curves we actually draw will not allow us the space to represent very many of its fluctuations.

Now let us look a little more closely at $f''(x)$. Suppose $f''(x_1) > 0$ at some point x_1. Then for small h, the difference quotient

$$\frac{f'(x_1+h) - f'(x_1)}{h}$$

is positive; if $h > 0$, $f'(x_1 + h) > f'(x_1)$, and if $h < 0$, $f'(x_1 + h) < f'(x_1)$. But $f'(x_1)$ is the slope of the tangent to the curve at x_1, so this tangent is given by the equation $y = f'(x_1)x + b$, where b is yet to be determined. But we must have $(x_1, f(x_1))$ on this line, or $f(x_1) = x_1 f'(x_1) + b$, and $b = f(x_1) - x_1 f'(x_1)$. Thus *the tangent line at* x_1 *is*

$$y = xf'(x_1) + (f(x_1) - x_1 f'(x_1)).$$

Now let x be in a neighborhood of x_1 so small that whenever $x_1 + h$ is in this neighborhood,

$$\frac{f'(x_1+h) - f'(x_1)}{h} > 0.$$

Then $f(x) - f(x_1) = f'(\xi)(x - x_1)$, where ξ is between x and x_1, therefore is surely in this neighborhood of x_1. If $x > x_1$, then $\xi > x_1$, and $f'(\xi) > f'(x_1)$ as we have seen above. Therefore

$$f(x) - f(x_1) > f'(x_1)(x - x_1),$$

or

$$f(x) > xf'(x_1) + (f(x_1) - x_1 f'(x_1)).$$

If $x < x_1$, then $f'(\xi) < f'(x_1)$, and

$$f(x_1) - f(x) < f'(x_1)(x_1 - x),$$

or

$$f(x) > xf'(x_1) + (f(x_1) - x_1 f'(x_1)).$$

In either case, the value of the original function is greater than the value of the function describing the tangent line, or *the curve lies above its tangent at* x_1 (meaning, of course, in a neighborhood of x_1). *In this case we say that the curve is concave upward at* x_1. Likewise, if $f''(x_1) < 0$, the curve lies *below* its tangent at x_1, in a neighborhood of x_1, and we call it *concave downward* at x_1. Thus the second derivative is related to the *curvature* of the curve; for example, if $f''(x_1)$ is large and positive, the curve bends sharply upward at x_1; if $f''(x_1)$ is negative but of small absolute value, the curve bends gradually downward at x_1.

Suppose now that the graph of $f(x)$ crosses its tangent at a point x_0. Then x_0 is called a point of inflection of $f(x)$. We have seen that if $f''(x_0) > 0$, the curve lies above its tangent in a neighborhood of x_0, and that if $f''(x_0) < 0$, it lies below the tangent. Therefore the only possible candidate for a point of inflection is a point x_0 where $f''(x_0) = 0$. Once again, not all such points need be points of inflection, as is seen from $y = x^4$.

If the tests we have mentioned fail to determine whether a point is a relative maximum or minimum, or a point of inflection, one should simply test that point by referring to the definitions of these terms. There are tests involving higher derivatives, but we shall not discuss them.

Now let us consider the general problem of tracing the graph of a function $y = f(x)$ with some attention to its critical points and in a relatively economical manner. First of all, we should observe what the domain of the function is, i.e., for what values of x it is defined. For instance, $f(x) = \frac{1}{x - 1}$ is defined for $x \neq 1$, $f(x) = \sqrt{x}$ for all $x \geqq 0$. We should also observe what its range is, i.e., what values $f(x)$ can take on. For example, the range of $y = \sin x$ is $-1 \leqq y \leqq 1$, while that of $y = \sqrt{x}$ is all $y \geqq 0$ and that of $y = x^3$ is all numbers. Next we should determine a few points of the graph, and one usually finds those where it crosses the axes. It will of course cross the y-axis when $x = 0$, $y = f(0)$. Sometimes finding where it crosses the x-axis, i.e., where $f(x) = 0$, is considerably harder, and must be estimated. The totality of such points are called the intercepts of the graph.

Then one should determine the turning points, i.e., relative maxima and minima and points of inflection. By the location of the maxima and minima and by the behavior of $f(x)$ for x very far positive and very far negative, it is possible to see in what direction the curve crosses its tangent at the points of inflection. It is also useful to determine in what regions $f''(x) > 0$ and $f''(x) < 0$, to see in what direction the curve is concave. In this process, some points may occur where the first or second derivative does not exist. They should be plotted separately, and special attention should be paid to the behavior of the curve near them.

Finally, one should look for the asymptotes of the curve, which are, roughly speaking, lines approached by the curve. Vertical asymptotes will often occur at points where the function is undefined; for example, the line $x = 1$ is a vertical asymptote to $y = \frac{1}{x - 1}$. Horizontal asymptotes will occur when $f(x)$ approaches a constant as x goes very far positive or very far negative. The line $y = 0$ is, for instance, a horizontal asymptote to $y = \frac{1}{x - 1}$, here both for x far negative and far positive. A suspected oblique asymptote can be treated by the test for horizontal asymptotes by taking its equation $y = mx + b$ and subtracting from $f(x)$ to get the new function $g(x) = f(x) - (mx + b)$. Then $y = mx + b$ is an asymptote for $f(x)$ if and only if $y = 0$ is an asymptote for $g(x)$.

By these considerations, one can make a sketch which will reveal the shape of the curve, since the behavior between the critical points can be plotted smoothly with attention only to the concavity in these regions. As an example, let us consider $f(x) = x^{2/3}(1 - x)$. This is defined for all x, becomes far negative as x becomes positive, far positive as x goes far negative, and takes on all values between. Thus its range is all numbers. We see that $f(0) = 0$, $f(1) = 0$. Thus the intercepts are $(0, 0)$ and $(1, 0)$. Now $f'(x) = \frac{2}{3} x^{-1/3}(1 - x) - x^{2/3}$, where this is defined. We see from the first term that $f'(0)$ is undefined. But for x either slightly positive or slightly negative, $f(x) > 0 = f(0)$. Therefore $x = 0$ is a relative minimum. In the remaining cases, setting $f'(x) = 0$ gives

$$0 = \frac{2}{3} x^{-1/3}(1 - x) - x^{2/3} = \frac{2}{3} x^{-1/3} - \frac{5}{3} x^{2/3} = \frac{2}{3} - \frac{5}{3} x,$$

or $x = \frac{2}{5}$ is the only solution of $f'(x) = 0$. Now

$$f''(x) = -\frac{2}{9} x^{-4/3}(1 - x) - \frac{2}{3} x^{-1/3} - \frac{2}{3} x^{-1/3}$$

$$= -\frac{2}{9} x^{-4/3} - \frac{10}{9} x^{-1/3} = -\frac{2}{9} x^{-4/3}(1 + 5x), \quad x \neq 0.$$

Thus $f''(\frac{2}{5}) < 0$, and $x = \frac{2}{5}$ is a relative maximum. If $f''(x) = 0$, then $x = -\frac{1}{5}$, and $f''(x) > 0$ for $x < -\frac{1}{5}$, $f''(x) < 0$ for $x > -\frac{1}{5}$, $x \neq 0$. Since the curvature changes sign at $-\frac{1}{5}$, $x = -\frac{1}{5}$ is a point of inflection, the graph is concave upward for $x < -\frac{1}{5}$, concave downward for $x > -\frac{1}{5}$. There are no apparent asymptotes. In the critical region, then, the curve looks like this:

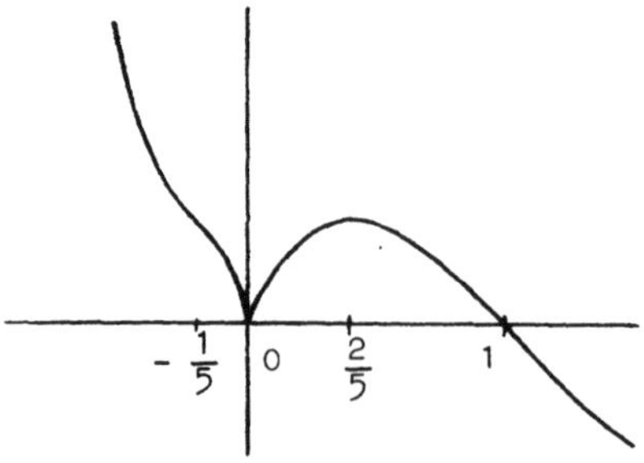

An equation for the tangent to the graph of $y = f(x)$ has been given above in the discussion of concavity. However, let us derive this in parametric form, as well as the equation of the normal, or perpendicular, to $y = f(x)$ at a point. A vector A parallel to the tangent line is found by taking any vector $A = (a_1, a_2)$

such that the ratio $\frac{a_2}{a_1}$ of its components is equal to the slope. At the point x_o, this slope is $f'(x_o)$, and we can take $a_1 = 1$, $a_2 = f'(x_o)$. The point of the curve is $(x_o, f(x_o))$, and so the parametric equation of the tangent line at $(x_o, f(x_o))$ is

$$(x, y) = (x_o, f(x_o)) + t(1, f'(x_o)).$$

Solving for t gives $t = x - x_o$; now $y = f(x_o) + tf'(x_o)$, or

$$y = f(x_o) + xf'(x_o) - x_of'(x_o) = xf'(x_o) + (f(x_o) - x_of'(x_o)),$$

the equation we have used before.

A vector parallel to the normal line at $(x_o, f(x_o))$ must be perpendicular to $(1, f'(x_o))$. Let (b_1, b_2) be such a vector. Then

$$b_1 + b_2f'(x_o) = 0, \quad \text{or} \quad \frac{b_1}{b_2} = - f'(x_o).$$

In particular, we can choose $b_2 = 1$, $b_1 = - f'(x_o)$. The normal line is then

$$(x, y) = (x_o, f(x_o)) + t(- f'(x_o), 1).$$

EXERCISES

Courant, p. 166, ex. 1-7, 14-18.
Oakley, p. 51, ex. 1-10.

10. The Definite Integral

Let $y = f(x)$ be a function continuous for $a \leq x \leq b$. We plot its graph; for instance, it might be as shown in the graph on the following page.

Now let us define a number which we shall call the (definite) integral from a to b of $f(x)$, and for which we write

$$\int_a^b f(x)dx, \quad \text{as follows:}$$

Consider the area bounded by the graph of $y = f(x)$, the x-axis, the line $x = a$ and the line $x = b$. If a part of this area lies above the x-axis, we shall consider it as positive. If a part lies below the x-axis, we

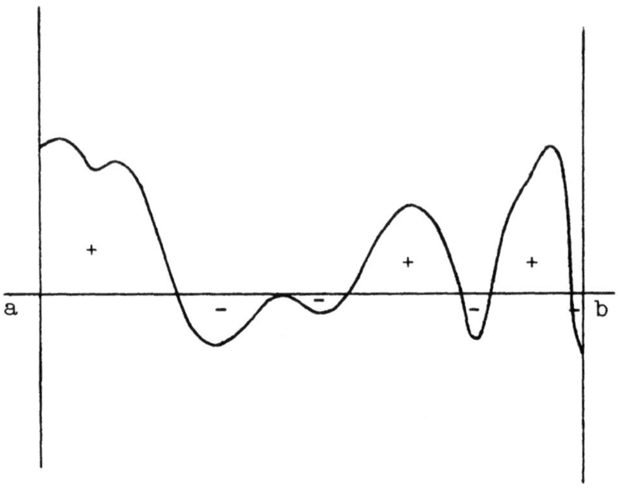

shall consider it as negative. Thus we have the arrangement of signs shown in the diagram. Then we define

$$\int_a^b f(x)dx$$

to be simply <u>the algebraic sum of these areas, taken with the corresponding</u> signs.

If $b \leqq a$, we define

$$\int_a^b f(x)dx = - \int_b^a f(x)dx,$$

since

$$\int_b^a f(x)dx$$

is then defined in the above manner. In particular, we have

$$\int_a^a f(x)dx = - \int_a^a f(x)dx,$$

or

$$\int_a^a f(x)dx = 0.$$

Also, if a, b, c are any numbers such that f(x) is continuous on the smallest interval containing them all, then

$$\int_a^c f(x)dx = \int_a^b f(x)dx + \int_b^c f(x)dx.$$

If $a \leqq b \leqq c$, this is obvious from the definition. It is left as an exercise to verify it in the other cases.

Now suppose x is in the interval $a \leqq t \leqq b$. Then

$$\int_a^x f(t)dt$$

has been defined, where we regard x as fixed for the moment, and denote our variable by t. Let us consider this as a function F(x). Then the difference quotient for F(x) is

$$\frac{F(x+h) - F(x)}{h} = \frac{\int_a^{x+h} f(t)dt - \int_a^x f(t)dt}{h} = \frac{1}{h} \cdot \int_x^{x+h} f(t)dt$$

by the above remark.

Now let M be the maximum of f(t) on the interval between x and x + h, m its minimum. Then we assert that if $h > 0$,

$$Mh \geqq \int_x^{x+h} f(t)dt \geqq mh, \quad \text{or} \quad M \geqq \frac{1}{h} \int_x^{x+h} f(t)dt \geqq m.$$

The first inequality is reversed if $h < 0$, but the second one is still valid. We shall show this in the case where $f(x) > 0$, $h > 0$. The idea for the remaining cases will then be clear. Because f(t) is continuous, we can assume h is small enough so that $f(t) > 0$ for all t between x and x + h. Then our statement amounts to saying that the area under the graph of f(t) between x and x + h lies between the area of a rectangle of width h whose height is the smallest value of f(t) in this interval and that of a rectangle of width h and height the maximum value of f(t) in the interval, as shown in the diagram on the following page.

But since f(t) is continuous at x, f(t) approaches f(x)

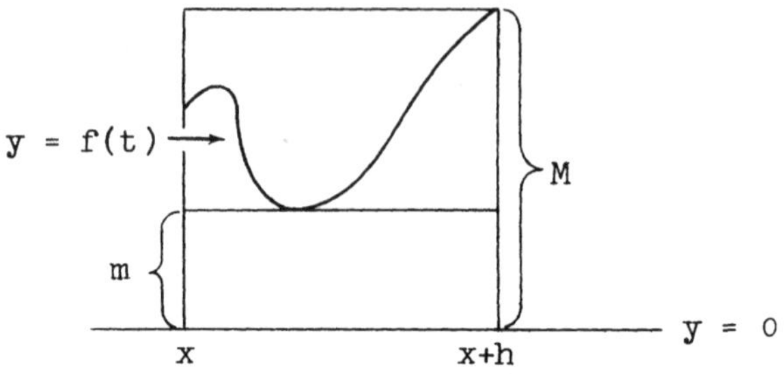

for all t between $f(x)$ and $f(x + h)$, as h approaches 0. Since M and m are values for $f(t)$, they also approach $f(x)$. Thus we have

$$f(x) = \lim_{h \to 0} M \geq \lim_{h \to 0} \frac{1}{h} \int_x^{x+h} f(t)dt \geq \lim_{h \to 0} m = f(x);$$

but then all these are equal, and $F'(x) = f(x)$.

Now we have a procedure for evaluating

$$\int_a^b f(t)dt.$$

First we find <u>any function</u> $G(x)$ such that $G'(x) = f(x)$ for $a \leq x \leq b$; such a function is called an <u>indefinite integral</u>, or an <u>antiderivative</u>, of $f(x)$. Then we know that

$$F(x) = \int_a^x f(t)dt$$

has the property that $F'(x) = f(x) = G'(x)$. Therefore $F(x) = G(x) + c$, where c is some constant. But $F(a) = 0$, so $c = - G(a)$. Then

$$\int_a^b f(t)dt = F(b) = G(b) + c = G(b) - G(a) .$$

This is the so-called <u>Fundamental Theorem of Integral Calculus</u>. By its use we can compute the definite integrals for many simple functions, therefore the areas defined by their graphs. We shall later develop some techniques which will extend the class of functions we can handle.

NOTE. Once again we have a <u>linear</u> operation; if $F(x)$ is an indefinite integral of $f(x)$, $G(x)$ one for $g(x)$, then

$$(F(x) + G(x))' = F'(x) + G'(x) = f(x) + g(x),$$

or $F(x) + G(x)$ is an indefinite integral for $f(x) + g(x)$. Likewise, $cF(x)$ is an indefinite integral for $cf(x)$. For definite integrals, with $F(x)$, $f(x)$, $G(x)$, $g(x)$ and c as above, we have

$$\int_a^b (f(t) + g(t))dt = (F(b) + G(b)) - (F(a) + G(a))$$

$$= (F(b) - F(a)) + (G(b) - G(a)) = \int_a^b f(t)dt + \int_a^b g(t)dt$$

and

$$\int_a^b cf(t)dt = cF(b) - cF(a) = c(F(b) - F(a)) = c \int_a^b f(t)dt.$$

Observe also that any two indefinite integrals for $f(x)$ differ only by a constant.

EXAMPLES

a)
$$\int_0^1 x^2 dx;$$

this is the area under the parabola $y = x^2$ between $x = 0$ and $x = 1$, as shaded in the figure. To find it, we must find a function $F(x)$ such that $F'(x) = x^2$. Now $(x^3)' = 3x^2$, so we use $\frac{1}{3}x^3$. This is then an indefinite integral $F(x)$, and the area is $F(1) - F(0) = \frac{1}{3} \cdot 1^3 - \frac{1}{3} \cdot 0^3 = \frac{1}{3}$.

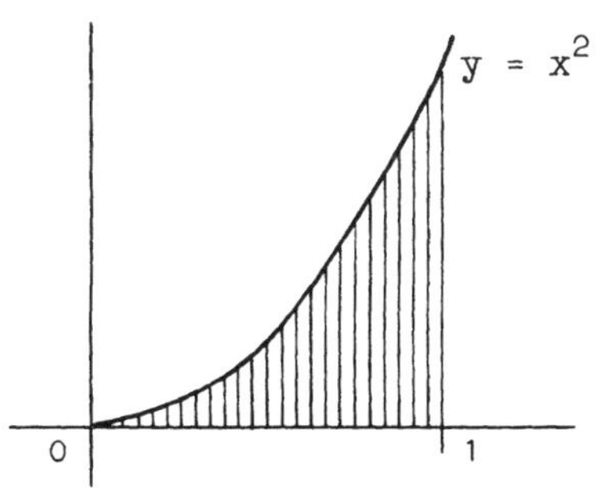

b)
$$\int_a^b c\,dx,$$

where c is a constant. An indefinite integral is cx, so our answer is $cb - ca = c(b - a)$. This is simply the area of the rectangle with sides $y = c$, $y = 0$, $x = a$, $x = b$; the sign is positive if $c > 0$, negative if $c < 0$.

c)

$$\int_a^b x^n dx,$$

n a positive integer. x^{n+1} has derivative $(n + 1)x^n$, so that an indefinite integral for x^n is $\frac{1}{n+1} x^{n+1}$. The result is simply $\frac{1}{n+1} (b^{n+1} - a^{n+1})$.

d) Using linearity, we can now find

$$\int_a^b f(x)dx$$

whenever $f(x)$ is a <u>polynomial</u>.

e) By examining the rules for differentiation, you will see that the indefinite integrals below are correct; we write

$$\int f(x)dx = F(x)$$

to mean simply that $F(x)$ is an indefinite integral of $f(x)$, or that $F'(x) = f(x)$:

1) $\int e^x dx = e^x$;

2) $\int \cos x \, dx = \sin x$;

3) $\int \sin x \, dx = - \cos x$;

4) $\int \sec^2 x \, dx = \tan x$;

5) $\int x^b dx = \frac{1}{b+1} x^{b+1}$, b a positive rational number;

6) $\int a^x dx = \frac{1}{\log a} \cdot a^x$, $a > 0$.

In rules 1) and 6), we are still discussing functions of which we have no real knowledge. Likewise we could have written $\int \frac{1}{x} dx = \log x$, but this and the rules above are to be regarded for the present only as pieces of information of use in developing technique. We shall in the next section <u>define</u> $\log x$ in terms of an integral.

Please observe that when we write

$$\int_a^b f(x)dx,$$

the expression is meaningless to us unless $f(x)$ is defined and continuous throughout the interval $a \leq x \leq b$. Thus

$$\int_0^4 \frac{1}{x}\,dx$$

is undefined, as is

$$\int_{-1}^1 \sqrt{x}\,dx .$$

The definite integral can be defined in such a way that more functions fall into its domain (though neither of these two does), but we shall not pursue this question further.

EXERCISES

1. Give a rule for finding an indefinite integral of a polynomial. Courant, p. 119, ex. 2; p. 144, ex. 11-18.

11. The Function log x

Let $x > 0$. Then the function $f(t) = \frac{1}{t}$ is defined and continuous on the interval between $t = 1$ and $t = x$. We define a function for $x > 0$, which we shall call $\log x$, by

$$\log x = \int_1^x \frac{1}{t}\,dt .$$

Obviously this function has one property which we have attributed to that rather indefinite function which we have previously written as $\log x$: namely, $(\log x)' = \frac{1}{x}$. The next basic property we prove is

$$\log (ab) = \log a + \log b .$$

But for any fixed a, consider $\log (ax)$ as a function of x. Then

$$(\log (ax))' = \frac{1}{ax} \cdot a = \frac{1}{x} = (\log x)' .$$

Therefore $\log (ax) = \log x + c$, c some constant. For $x = 1$, we have

$$\log a = \log 1 + c.$$

Since

$$\log 1 = 0 \quad \left(= \int_1^1 \frac{1}{t}\, dt \right),$$

$c = \log a$, and $\log (ax) = \log x + \log a$. This is what we wished to prove.

Now let $a > 1$; then $\log a \leqq a -$ for this is the area of the larger rectangle in the diagram at the left. The smaller rectangle has area $(a - 1) \cdot \frac{1}{a} \leqq \log a$. Thus we have for $a > 1$,

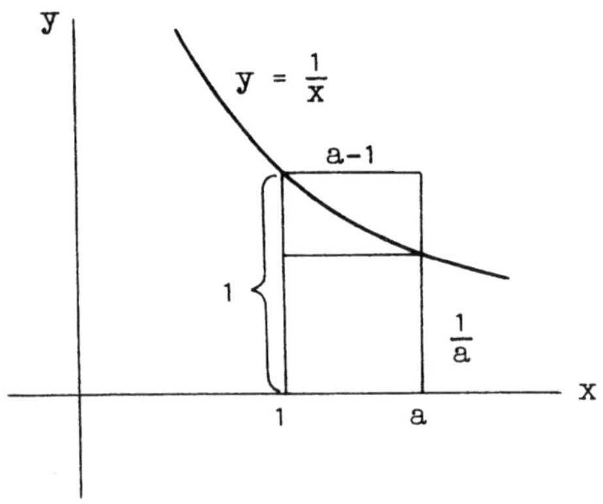

$$a - 1 \geqq \log a \geqq \frac{a - 1}{a} = 1 - \frac{1}{a}.$$

This is an equality when $a = 1$, so is true whenever $a \geqq 1$. Now if $a < 1$, then $\frac{1}{a} > 1$, and we have

$$\frac{1}{a} - 1 \geqq \log \frac{1}{a} \geqq 1 - \frac{1}{1/a}$$

by the above. Now

$$\log a + \log \frac{1}{a} = \log 1 = 0,$$

so

$$\log \frac{1}{a} = - \log a.$$

This gives

$$\frac{1}{a} - 1 \geqq - \log a \geqq 1 - a,$$

or, reversing signs

(*) $$a - 1 \geqq \log a \geqq 1 - \frac{1}{a}.$$

Thus we have proved the inequality (*) for all $a > 0$.

Let us use this inequality to derive some properties of $\log x$. Using $a = 1 + \frac{1}{n}$, n a positive integer, we find

$$\frac{1}{n} \geqq \log \left(1 + \frac{1}{n}\right) \geqq 1 - \frac{1}{1 + \frac{1}{n}} = 1 - \frac{n}{n + 1} = \frac{1}{n + 1}.$$

Since $n(\log a) = \log (a^n)$ (prove by induction), we have

$$1 = \frac{n}{n} \geqq n \log \left(1 + \frac{1}{n}\right) = \log \left(1 + \frac{1}{n}\right)^n \geqq \frac{n}{n+1} .$$

Now as n becomes large, $\frac{n}{n+1}$ approaches 1. Therefore $\log \left(1 + \frac{1}{n}\right)^n$ is between 1 and a quantity which approaches 1, and hence must itself approach 1. We write this as

$$\lim_{n \to \infty} \log \left(1 + \frac{1}{n}\right)^n = 1.$$

Next consider the expression $\log (2^n) = n \log 2 \geqq n(1 - \frac{1}{2}) = \frac{n}{2}$. As n becomes large, $\frac{n}{2}$ also becomes large, and consequently $\log (2^n)$ becomes as large as we wish. Similarly, $\log(2^{-n}) \leqq -\frac{n}{2}$, and as n becomes arbitrarily large, $\log(2^{-n})$ becomes arbitrarily small (i.e., far negative). Thus we see that $\log x$ takes on arbitrarily large and arbitrarily small values.

$\log x$ is <u>continuous</u> at every $x > 0$, since it is differentiable. Moreover, it is <u>strictly increasing</u>. If $x_1 < 1$, $x_2 \geqq 1$, it is clear that $\log x_1 < 0 \leqq \log x_2$. If $1 \leqq x_1 < x_2$, then the area under $y = \frac{1}{x}$ from 1 to x_1 is evidently less than that from 1 to x_2, so again $\log x_1 < \log x_2$. Finally, if $x_1 < x_2 < 1$, the area from x_1 to 1 is greater than that from x_2 to 1; but the logarithms are these areas taken with a negative sign, giving once more $\log x_1 < \log x_2$. Thus we have proved the assertion.

12. <u>Inverse Functions and the Exponential</u>

Consider the following question: Given $y = f(x)$ for $a \leqq x \leqq b$, when can we turn this relationship around, i.e., when does this mean the same thing as $x = g(y)$ for some function $g(y)$? Thus we want to have $y = f(x)$ if and only if $x = g(y)$. Observe that our notation means that $g(y)$ is to assign to each y of a certain domain <u>exactly one</u> value of x. Therefore if we have $y_0 = f(x_1)$, $y_0 = f(x_2)$, where $x_1 \neq x_2$, we cannot define such a function $g(y)$ at y_0. So we must have $f(x_1) \neq f(x_2)$ if $x_1 \neq x_2$. We already know two important classes of functions which have this property: those which are <u>strictly increasing</u> and those which are <u>strictly decreasing</u>.

The function whose graph is shown on the following page is strictly increasing for $0 \leqq x \leqq 1$. The function $g(y)$, which we call the <u>inverse function</u>, is given by

$$g(y) = y, \quad 0 \leqq y < \frac{1}{2}$$
$$g(y) = y - 1, \quad \frac{3}{2} \leqq y \leqq 2.$$

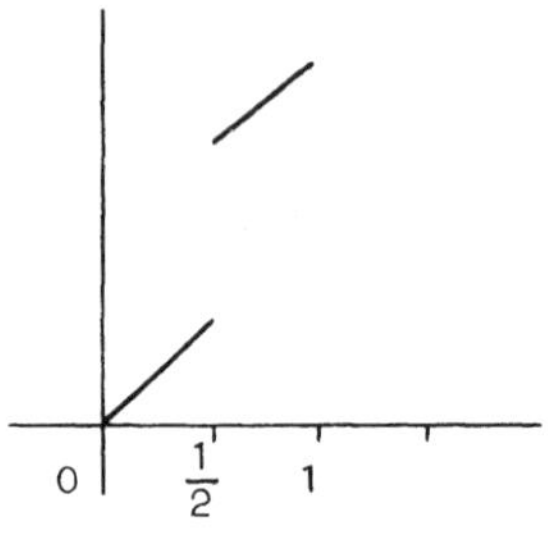

$$\begin{cases} f(x) = x, & 0 \leqq x < \frac{1}{2} \\ f(x) = 1 + x, & \frac{1}{2} \leqq x \leqq 1. \end{cases}$$

We prefer, however, to restrict our attention to functions which can be defined throughout an entire interval, rather than on a broken interval as in this case. We could eliminate the "jump" in $f(x)$ by requiring that it be continuous. This requirement gives us considerable information about the inverse function, because of the following theorem which will be proved later:

> THEOREM. Let $f(x)$ be continuous for $a \leqq x \leqq b$. Let $f(a) < f(b)$, and let $f(a) \leqq y_0 \leqq f(b)$. Then there is an x_0, $a \leqq x_0 \leqq b$, such that $y_0 = f(x_0)$.

Now let $f(x)$ be continuous and strictly increasing for $a \leqq x \leqq b$. (We could also assume $f(x)$ continuous and strictly decreasing.) Then $f(a) < f(b)$, and if $a \leqq x \leqq b$, then $f(a) \leqq f(x) \leqq f(b)$. Moreover, if $f(a) \leqq y \leqq f(b)$, there is an x, $a \leqq x \leqq b$, such that $y = f(x)$. We can then define the inverse function $g(y)$ on the whole interval $f(a) \leqq y \leqq f(b)$:

$$x = g(y) \text{ is to mean simply that } y = f(x).$$

Once we have given a precise definition of continuity, it will be easy to see that $g(y)$ is continuous. It is clearly strictly increasing.

For the present, let us assume that $f(x)$ is also differentiable; we wish to express $g'(y)$ in terms of $f'(x)$. Let us also make the somewhat stronger assumption that $f'(x) > 0$ for $a \leqq x \leqq b$. (For instance, $y = x^3$ is strictly increasing for $-1 \leqq x \leqq 1$, but $y' = 0$ at 0. However, we have seen that if $f'(x) > 0$ for all x of an interval, then $f(x)$ is strictly increasing there.) Then let $f(a) < y < f(b)$, and consider

$$\frac{g(y+h) - g(y)}{h} .$$

Now $y = f(x)$ and $y + h = f(x + k)$ for some x and $x + k$, $a \leqq x \leqq b$,

$a \leqq x + k \leqq b$. Therefore the difference quotient for $g(y)$ is

$$\frac{(x+k) - x}{f(x+k) - f(x)} = \frac{k}{f(x+k) - f(x)} = \frac{1}{\frac{f(x+k) - f(x)}{k}} .$$

By our remark that $g(y)$ is continuous, the quantity $g(y + h) - g(y) = k$ goes to 0 as $h \longrightarrow 0$. But as $k \longrightarrow 0$,

$$\frac{f(x+k) - f(x)}{k}$$

goes to $f'(x)$. Therefore we have

$$g'(y) = \lim_{h \longrightarrow 0} \frac{g(y+h) - g(y)}{h} = \frac{1}{f'(x)} ,$$

where $y = f(x)$, $x = g(y)$. Thus we have the rule $g'(y) = \frac{1}{f'(x)}$ for differentiating inverse functions.

We have seen that for $x > 0$, $y = \log x$ is continuous and strictly increasing; moreover, its derivative $\frac{1}{x} > 0$ for $x > 0$. Therefore it has an inverse function, which we denote temporarily by $E(y)$ and later by e^y, after we have shown that its properties entitle it to this notation. Thus we define $E(y)$, for <u>any</u> number y, by

$$x = E(y) \text{ if and only if } y = \log x.$$

If $y = \log x$, then

$$E'(y) = \frac{1}{(\log x)'} = \frac{1}{1/x} = x = E(y),$$

or $E'(y) = E(y)$. We have already attributed this property to e^y. We also see that $E(y) > 0$ for all y, and that $E(0) = 1$. When y is far negative, $E(y)$ is close to 0, and when y is large and positive, so is $E(y)$. Now $E(y_1 + y_2)$ is that number x whose log is $y_1 + y_2$. But if $\log x_1 = y_1$, $\log x_2 = y_2$, then $\log(x_1x_2) = y_1 + y_2$, as we have seen. Hence the desired number is x_1x_2, i.e., $E(y_1)E(y_2)$, and we have the relation

$$E(y_1 + y_2) = E(y_1)E(y_2).$$

It follows that for any positive integer n, $E(n) = E(1)^n$. Thus we have recovered some of the exponential-like properties of this function. Let us <u>define</u> a number e by $e = E(1)$, so that e is that number such that $\log e = 1$. Then we see that for positive integers n, $E(n) = e^n$. We agree to extend this notation to all y, and to write $e^y = E(y)$. Thus $e^y > 0$ for all y, $e^0 = 1$,

$$e^{y_1+y_2} = e^{y_1}e^{y_2}, \quad (e^y)' = e^y.$$

As indicated before, we can now define a^b for $a > 0$ by $a^b = e^{b \log a}$. Then

$$a^{b_1+b_2} = e^{b_1\log a + b_2\log a} = e^{b_1\log a}e^{b_2\log a} = a^{b_1}a^{b_2},$$

and similarly we can show that

$$(a^b)^c = a^{bc}; \quad a^0 = 1; \quad a^1 = a; \quad a^n = \overbrace{a \cdot a \cdot \ldots a}^{n},$$

n a positive integer;

$$a^{-n} = \frac{1}{a^n}.$$

It is of interest to introduce some other inverse functions, notably those of the trigonometric functions. It is obvious from their periodic behavior that these functions are neither strictly increasing nor strictly decreasing; but all this means is that we cannot hope to define an inverse function whose values (whose range, in more technical language) cover the whole x-axis. For instance, the function $y = \sin x$ is strictly increasing for

$$-\frac{\pi}{2} \leqq x \leqq \frac{\pi}{2},$$

and $y' > 0$ except at the endpoints. The values run from $y = -1$ to $y = 1$. Hence we can define an inverse function to $y = \sin x$, written as $x = \arcsin y$, for $-1 \leqq y \leqq 1$. Its values lie between $x = -\frac{\pi}{2}$ and $x = \frac{\pi}{2}$. Its derivative is

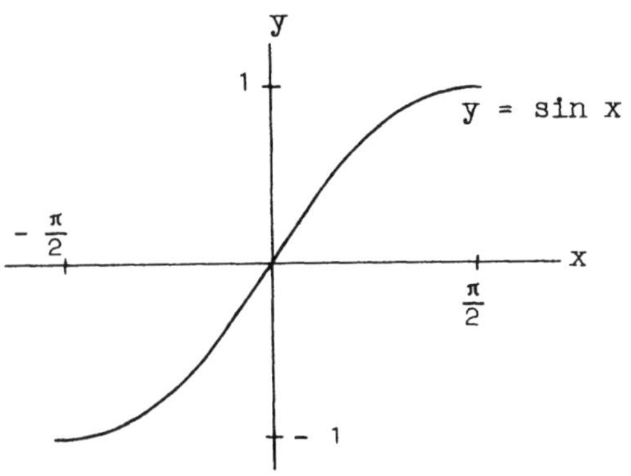

$$\frac{1}{(\sin x)'} = \frac{1}{\cos x}.$$

We should like to express this in terms of y. Since $\sin^2 x + \cos^2 x = 1$, and since $\cos x > 0$ for $-\frac{\pi}{2} < x < \frac{\pi}{2}$, we have

$$\cos x = \sqrt{1 - \sin^2 x} = \sqrt{1 - y^2} .$$

Thus

$$(\arcsin y)' = \frac{1}{\sqrt{1-y^2}} , \quad -1 < y < 1 .$$

Likewise, $x = \arccos y$ is defined for $-1 \leqq y \leqq 1$ and takes on values between $x = 0$ and $x = \pi$. (Here we use the inverse function of a strictly decreasing function.)

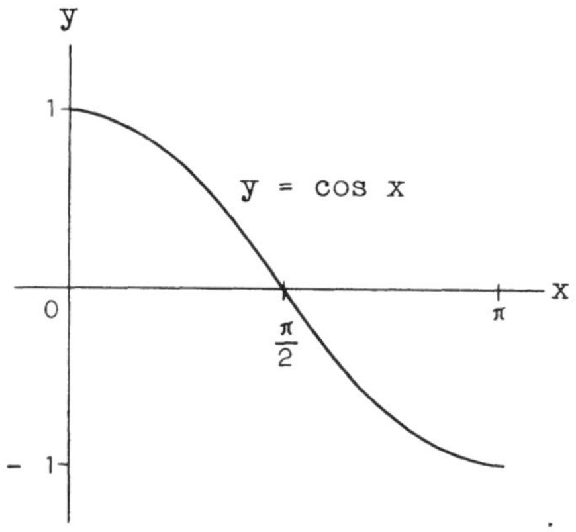

By the above process,

$$(\arccos y)' = \frac{-1}{\sqrt{1-y^2}} , \quad -1 < y < 1.$$

$x = \arctan y$ is similarly defined, in this case for all y, using the fact that $y = \tan x$ is strictly increasing for

$$-\frac{\pi}{2} < x \frac{\pi}{2} .$$

This is then the range of values of $x = \arctan y$. As before,

$$(\arctan y)' = \frac{1}{1 + y^2}$$

As a final remark, let us add that the only functions of the form x^a for which we had a really adequate definition at the beginning of this course were those for which a was an integer. For n a positive integer, the function $y = x^n$ is continuous and strictly increasing for $x \geqq 0$ if n is even, for all x if n is odd. Thus we have the inverse function $x = y^{1/n}$, defined as above: $x = y^{1/n}$ if and only if $y = x^n$. Finally, we can define

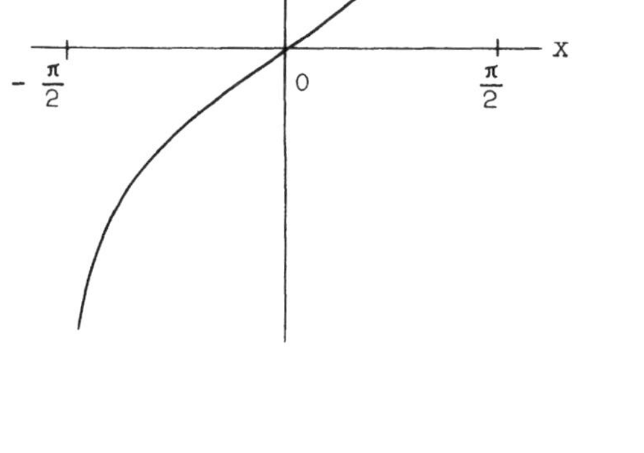

$$x^{m/n} = \left(x^{1/n}\right)^m = \left(x^m\right)^{1/n} ,$$

for m and n positive; and for b a negative rational number, define

$$x^b = \frac{1}{x^{-b}} , \quad x \neq 0.$$

Thus x^b is defined for b <u>rational</u>; we have not yet defined x^a for arbitrary a. This is accomplished by using $x^a = e^{a \log x}$. Since this definition coincides with the one already given for a rational, we may use it as the general definition. Then we have

$$(x^a)' = \left(e^{a \log x}\right)' = e^{a \log x} \cdot \frac{a}{x} = e^{a \log x} \cdot a \cdot e^{-\log x}$$

$$= a \cdot e^{(a-1)\log x} = a\, x^{a-1},$$

i.e., $(x^a)' = a\, x^{a-1}$ for all a.

13. Techniques of Integration

A. <u>Integration by parts</u>. Recall the rule $(f(x)g(x))' = f'(x)g(x) + f(x)g'(x)$. Now suppose that we try to find the indefinite integral

$$\int h(x)\, dx,$$

where h(x) can be written in the form $f(x)g'(x)$. It may perhaps be easier to find

$$\int g(x)f'(x)\, dx$$

than to find

$$\int h(x)\, dx.$$

By linearity,

$$\int (f(x)g(x))'\, dx = \int f'(x)g(x)\, dx + \int f(x)g'(x)\, dx,$$

or

$$f(x)g(x) = \int f'(x)g(x)\, dx + \int h(x)\, dx,$$

i.e.,

$$\int h(x)\, dx = f(x)g(x) - \int f'(x)g(x)\, dx.$$

Thus an indefinite integral of $h(x)$ is found by subtracting from $f(x)g(x)$ an indefinite integral of $f'(x)g(x)$.

EXAMPLES

1) $\int x \sin x \, dx$. Let $f(x) = x$, $g(x) = -\cos x$. Then $f'(x) = 1$, $g'(x) = \sin x$, and substituting gives

$$\int x \sin x \, dx = -x \cos x - \int (-\cos x) \, dx$$

$$= -x \cos x + \int \cos x \, dx = -x \cos x + \sin x.$$

2) $\int \log x \, dx$. Let $f(x) = \log x$, $g(x) = x$. Then $f'(x) = \frac{1}{x}$, $g'(x) = 1$, and

$$\int \log x \, dx = x \log x - \int \frac{1}{x} \cdot x \, dx = x \log x - x.$$

Sometimes it is necessary to perform the procedure several times. It often happens that the original integral will recur with a coefficient other than 1, in which case we can solve for it. For instance, consider $\int e^x \sin x \, dx$. Let $f(x) = e^x$, $g(x) = -\cos x$. Then

$$\int e^x \sin x \, dx = -e^x \cos x + \int e^x \cos x \, dx.$$

In $\int e^x \cos x \, dx$, let $h(x) = e^x$, $k(x) = \sin x$. Then $e^x \cos x = h(x)k'(x)$, and we have

$$\int e^x \cos x \, dx = e^x \sin x - \int e^x \sin x \, dx.$$

Thus

$$\int e^x \sin x \, dx = -e^x \cos x + e^x \sin x - \int e^x \sin x \, dx.$$

Solving for $\int e^x \sin x \, dx$, we find

$$\int e^x \sin x \, dx = \frac{e^x}{2} (\sin x - \cos x).$$

(NOTE: If at the second stage we had used $e^x \cos x = h(x)k'(x)$ with $h(x) = \cos x$, $k(x) = e^x$, the resulting equation would have been a trivial identity; thus if such an equation results, one should attempt a different decomposition.)

B. Integration by substitution. Certain integrals become much simpler if one variable is replaced by a new one. Here one uses in place of the old variable a new one which is, in the range of integration, a strictly increasing or decreasing differentiable function of the old one. If u is the new variable and x the old one, we then have $u'(x) = \frac{1}{x'(u)}$, from the differentiation of inverse functions.

Suppose we are to find an indefinite integral $\int f(x)\,dx$. That is, we are to find a function $F(x)$ such that $F'(x) = f(x)$. It may be that x is a strictly increasing or decreasing differentiable function of some new variable u, say $x = x(u)$, such that $f(x(u))$, which we shall call $g(u)$, is a simpler function to integrate. Then we might try to find a function $G(u)$ such that $G'(u) = g(u)$. If $u(x)$ is the inverse function to $x(u)$, we might try $H(x) = G(u(x))$ as an answer to our original problem. But

$$H'(x) = G'(u(x)) \cdot u'(x) = g(u(x)) \cdot u'(x) = f(x) \cdot u'(x),$$

which differs from $f(x)$ by the factor $u'(x) = \frac{1}{x'(u)}$. Therefore let us try instead to find a function $G(u)$ such that $G'(u) = g(u) \cdot x'(u)$. Then set $F(x) = G(u(x))$. We have then

$$F'(x) = G'(u(x)) \cdot u'(x) = g(u) \cdot x'(u) \cdot u'(x) = g(u) = f(x),$$

or $F(x)$ is a function of the desired type.

There is a formal device for remembering this calculation, as follows: In $\int f(x)\,dx$, write $dx = \frac{dx}{du} \cdot du$. If $f(x) = g(u)$, then

$$\int f(x)\,dx = \int g(u) \cdot \frac{dx}{du} \cdot du\ .$$

Of course, the answer on the right is a function of u, for which we must then substitute in terms of x. Note that the integral on the right is just $\int g(u)x'(u)\,du$, or our $G(u)$. Several examples will illustrate the technique.

1) $$\int \frac{e^x}{1 + e^{2x}}\,dx\ .$$

Let $u = e^x$, $x = \log u$. Then $x'(u) = \frac{1}{u}$. We must then find

$$\int \frac{u}{1 + u^2} \cdot \frac{1}{u} \cdot du = \int \frac{1}{1 + u^2}\,du = \arctan u\ .$$

Our desired indefinite integral is thus $\arctan(e^x)$.

2)
$$\int \frac{x^3}{\sqrt{1 - x^2}}\, dx .$$

Let $x = \sin u$ (this is possible, since

$$\frac{1}{\sqrt{1 - x^2}}$$

is only defined if $-1 < x < 1$). Then $u = \arcsin x$, $x'(u) = \cos u$, and $\sqrt{1 - x^2} = \cos u$, since $-\frac{\pi}{2} < u < \frac{\pi}{2}$ when u is in the range of $\arcsin x$. Thus we must find

$$\int \frac{\sin^3 u}{\cos u} \cdot \cos u\, du = \int \sin^3 u\, du$$

$$= \int \sin u(1 - \cos^2 u)\, du = - \cos u - \int \cos^2 u \sin u\, du.$$

Now let $v = \cos u$. Then $v'(u) = - \sin u$, $u'(v) = - \frac{1}{\sin u}$, and we have, at least for $0 < u < \frac{\pi}{2}$ (**),

$$\int \cos^2 u \sin u\, du = \int v^2 \cdot \sin u \cdot - \frac{1}{\sin u}\, dv = - \frac{1}{3} v^3 = - \frac{1}{3} \cos^3 u.$$

Hence

$$\int \sin^3 u\, du = - \cos u + \frac{\cos^3 u}{3},$$

and

$$\int \frac{x^3}{\sqrt{1 - x^2}}\, dx = \frac{\cos^3(\arcsin x)}{3} - \cos(\arcsin x) = \frac{(1-x^2)^{3/2}}{3} - \sqrt{1 - x^2} .$$

Since this function does have

$$\frac{x^3}{\sqrt{1 - x^2}}$$

as derivative whenever $-1 < x < 1$, we have not lost anything at those points in the process, such as (**), where we might have done so.

C. Partial fractions. Suppose we want to find an indefinite integral of $\frac{f(x)}{g(x)}$, where $f(x)$ and $g(x)$ are polynomials (such functions are called rational functions). We may assume that the degree of $f(x)$

(the highest power of x appearing) is less than the degree of $g(x)$; for if not, we can divide $f(x)$ by $g(x)$ and get

$$\frac{f(x)}{g(x)} = P(x) + \frac{h(x)}{g(x)},$$

where $P(x)$ is a polynomial and $h(x)$, the remainder in the division of $f(x)$ by $g(x)$, is of degree less than that of $g(x)$. Suppose also that we can factor $g(x)$ into products of polynomials of the forms $(x - a)$ and $x^2 + bx + c$, $b^2 - 4c < 0$. (It is always possible to do this, although it is usually difficult to determine the a's, b's and c's.) Once this factorization is accomplished, we begin a procedure which enables us to integrate $\frac{f(x)}{g(x)}$.

What we do is to separate the factors of the above forms in $g(x)$ and try to write $\frac{f(x)}{g(x)}$ as a sum of quotients of polynomials. The denominators will be the separate factors of $g(x)$ and the numerators will have degrees less than those of the denominators. If a factor $x - a$ or $x^2 + bx + c$ divides $g(x)$ more than once, say n times, we have to allow for n terms, one having the factor as denominator to each of the powers $1, 2, \ldots, n$. In each of these terms, it is still sufficient to assume that the numerator is a constant or of degree one (in case the denominator is a power of $x^2 + bx + c$). For instance, if we had something of the form

$$\frac{dx^3 + ex^2 + fx + g}{(x^2+bx+c)^2} = \frac{dx^3 + ex^2}{(x^2+bx+c)^2} + \frac{fx + g}{(x^2+bx+c)^2},$$

we could divide $dx^3 + ex^2$ by $x^2 + bx + c$ to get a quotient $P(x)$ of degree 1 and a remainder $R(x)$ of degree at most 1. Then

$$\frac{dx^3 + ex^2}{(x^2+bx+c)^2} = \frac{P(x)}{x^2 + bx + c} + \frac{R(x)}{(x^2+bx+c)^2},$$

so that the expression is reduced to the form which we shall consider.

Let $g(x) = (x - a_1)^{m_1} \ldots (x - a_r)^{m_r}(x^2 + b_1x + c_1)^{n_1} \ldots (x^2 + b_sx + c_s)^{n_s}$. Let $n = m_1 + \ldots + m_r + 2n_1 + \ldots + 2n_s$. Then $f(x)$ has degree less than n, and we can write $f(x) = d_{n-1}x^{n-1} + d_{n-2}x^{n-2} + \ldots + d_1x + d_o$, where some of the d's may be 0. Now we try to write

$$\frac{f(x)}{g(x)} = \frac{A_{1,1}}{x - a_1} + \frac{A_{1,2}}{(x-a_1)^2} + \ldots + \frac{A_{1,m_1}}{(x-a_1)^{m_1}} + \frac{A_{2,1}}{x - a_2} + \ldots + \frac{A_{r,m_r}}{(x-a_r)^{m_r}}$$
$$+ \frac{B_{1,1}x + C_{1,1}}{x^2 + b_1x + c_1} + \ldots + \frac{B_{1,n_1}x + C_{1,n_1}}{(x^2+b_1x+c_1)^{n_1}} + \ldots + \frac{B_{s,n_s}x + C_{s,n_s}}{(x^2+b_sx+c_s)^{n_s}}.$$

Here the A, B, C are unknowns, and there are $m_1 + \cdots + m_2 + 2n_1 + \cdots + 2n_s = n$ of them in all. If we put the right-hand side of the equation over a common denominator, this denominator is $g(x)$. Therefore the numerator must be $f(x)$; so the constant term of the numerator must be d_o, the coefficient of x must be d_1, etc. This gives us n equations in the n unknowns A, B, C, which we solve. Substituting these solutions in the right-hand side of the equation, it remains only to integrate that expression term-wise. Now

$$\int \frac{1}{x-a}\,dx = \log(x-a); \quad \int \frac{1}{(x-a)^n}\,dx = \frac{-1}{n-1}\cdot\frac{1}{(x-a)^{n-1}}, \qquad n > 1.$$

Since $x^2 + bx + c = (x + \frac{b}{2})^2 + (c - \frac{b^2}{4})$, $c - \frac{b^2}{4} > 0$, we can substitute $x + \frac{b}{2}$ for x and assume that $x^2 + bx + c$ is simply of the form $x^2 + c$, $c > 0$.

$$\int \frac{x}{x^2+c}\,dx = \frac{1}{2}\log(x^2+c); \quad \int \frac{1}{x^2+c}\,dx = \frac{1}{\sqrt{c}}\arctan\left(\frac{x}{\sqrt{c}}\right);$$

$$\int \frac{x}{(x^2+c)^n}\,dx = \frac{-1}{2(n-1)}\cdot\frac{1}{(x^2+c)^{n-1}}, \qquad n > 1;$$

and

$$\int \frac{1}{(x^2+c)^n}\,dx$$

can be handled by the substitution $x = \sqrt{c}\tan u$, or $u = \arctan\left(\frac{x}{\sqrt{c}}\right)$.

EXAMPLES

1) $$\int \frac{1}{(x-2)(x+3)}\,dx.$$

Then

$$\frac{1}{(x-2)(x+3)} = \frac{A}{x-2} + \frac{B}{x+3},$$

or

$$\frac{(A+B)x + (3A-2B)}{(x-2)(x+3)} = \frac{1}{(x-2)(x+3)}.$$

Then $A + B = 0$, $3A - 2B = 1$, $A = \frac{1}{5}$, $B = -\frac{1}{5}$, and

$$\int \frac{1}{(x-2)(x+3)}\,dx = \frac{1}{5}\log(x-2) - \frac{1}{5}\log(x+3).$$

2)
$$\int \frac{4x+3}{(x-1)(x^2+2)^2}\,dx\,.$$

Write

$$\frac{4x+3}{(x-1)(x^2+2)^2} = \frac{A}{x-1} + \frac{Bx+C}{x^2+2} + \frac{Dx+E}{(x^2+2)^2}\,.$$

We get the equations

$A + B = 0$ (coefficient of x^4)
$-B + C = 0$ (coefficient of x^3)
$4A + 2B - C + D = 0$ (coefficient of x^2)
$-2B + 2C - D + E = 4$ (coefficient of x)
$4A - 2C - E = 3$ (constant term).

They give:

$$A = \frac{7}{9},\quad B = -\frac{7}{9} = C,\quad D = -\frac{7}{3},\quad E = \frac{5}{3}\,.$$

Thus we must find

$$\frac{7}{9}\int \frac{1}{x-1}\,dx - \frac{7}{9}\int \frac{x}{x^2+2}\,dx - \frac{7}{9}\int \frac{1}{x^2+2}\,dx - \frac{7}{3}$$

$$\int \frac{x}{(x^2+2)^2}\,dx + \frac{5}{3}\int \frac{1}{(x^2+2)^2}\,dx = \frac{7}{9}\log(x-1) - \frac{7}{18}\log(x^2+2)$$

$$-\frac{7}{9\sqrt{2}}\arctan\frac{x}{\sqrt{2}} + \frac{7}{6}\cdot\frac{1}{x^2+2} + \frac{5}{3}\int \frac{1}{(x^2+2)^2}\,dx\,.$$

In the last integral, let $x = \sqrt{2}\tan u$. Then $x'(u) = \sqrt{2}\sec^2 u$, and we must find

$$\int \frac{1}{4\sec^4 u}\cdot\sqrt{2}\sec^2 u\,du = \int \frac{1}{(x^2+2)^2}\,dx\,,$$

or

$$\frac{1}{2\sqrt{2}}\int \cos^2 u\,du\,.$$

Now $\cos 2u = 2\cos^2 u - 1$, or $\cos^2 u = \frac{1}{2}(1 + \cos 2u)$. Thus we must find

$$\frac{1}{4\sqrt{2}} \int (1 + \cos 2u)\, du = \frac{1}{4\sqrt{2}} u + \frac{1}{8\sqrt{2}} \sin 2u$$

$$= \frac{1}{4\sqrt{2}} \arctan \frac{x}{\sqrt{2}} + \frac{1}{4\sqrt{2}} \frac{\tan u}{\sec^2 u} = \frac{1}{4\sqrt{2}} \arctan \frac{x}{\sqrt{2}} + \frac{1}{4\sqrt{2}} \cdot \frac{\frac{x}{\sqrt{2}}}{\frac{x^2}{2} + 1}$$

$$= \frac{1}{4\sqrt{2}} \arctan \frac{x}{\sqrt{2}} + \frac{1}{4} \frac{x}{x^2 + 2} .$$

The complete answer is then

$$\frac{7}{9} \log(x - 1) - \frac{7}{18} \log(x^2 + 2) - \frac{7}{9\sqrt{2}} \arctan \frac{x}{\sqrt{2}} + \frac{7}{6} \cdot \frac{x}{x^2 + 2}$$

$$+ \frac{5}{12\sqrt{2}} \arctan \frac{x}{\sqrt{2}} + \frac{5}{12} \frac{x}{x^2 + 2} .$$

By a suitable substitution, one can reduce integrals of rational functions in $\sin x$ and $\cos x$ to the problem of integrating rational functions of a new variable u. The trick is to let $u = \tan \frac{x}{2}$. Then $x = 2 \arctan u$, and

$$x'(u) = \frac{2}{1 + u^2} .$$

Also,

$$\sin x = 2 \sin \frac{x}{2} \cos \frac{x}{2} = \frac{2 \tan \frac{x}{2}}{\sec^2 \frac{x}{2}} = \frac{2u}{1 + u^2} ,$$

and

$$\cos x = 2 \cos^2 \frac{x}{2} - 1 = \frac{2}{1 + \tan^2 \frac{x}{2}} - 1 = \frac{2}{1 + u^2} - 1 = \frac{1 - u^2}{1 + u^2} .$$

Thus, for instance,

$$\int \frac{1}{\sin x + \cos x}\, dx = \int \frac{1}{\frac{2u}{1 + u^2} + \frac{1 - u^2}{1 + u^2}} \cdot \frac{2}{1 + u^2}\, du$$

$$= \int \frac{2}{2u + 1 - u^2}\, du .$$

Now $2u + 1 - u^2 = -(u - (1 + \sqrt{2}))(u - (1 - \sqrt{2}))$, and we can apply the technique of partial fractions to complete the work.

EXERCISES

Courant: p. 225, ex. 1-11; p. 234, ex. 1-14.

Oakley: p. 100, ex. 1-12.

Sherwood and Taylor, Calculus, 1st ed.: pp. 202, 206, 215, 218, 222, 229, 231, 377, 380.

14. Curves in Parametric Form. Foundations of Mechanics

The illustration below is an example of something which we customarily call a curve, but which cannot be described as the graph of a function; for there are points where several y-values correspond to one value of x.

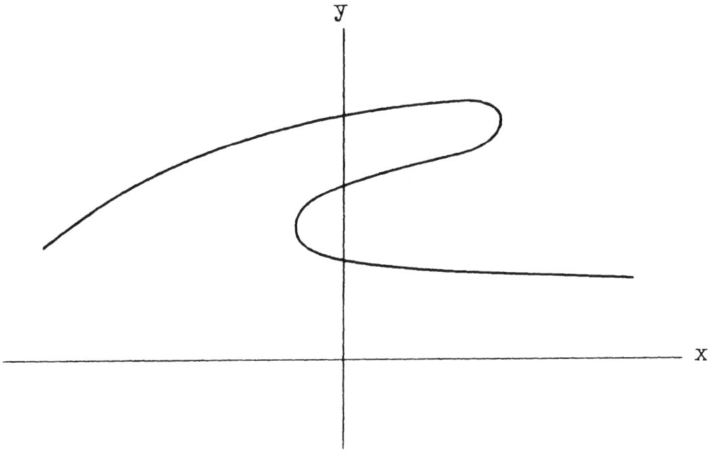

Then how can we define a curve so as to take in this case? When we drew the curve, we started with the pencil at some point and moved it. After t seconds the pencil was at a new point $(x(t), y(t))$ of the curve, and any point of the curve could be reached after some time (perhaps including travel backward in time). Now $x(t)$ is a function of t, namely, "the x-coordinate of the pencil at time t", and likewise $y(t)$ is a function of t. The fact that we draw the curve in one continuous motion is expressed by saying that $x(t)$ and $y(t)$ are continuous functions. Thus we are led to make the following definition:

A continuous curve in n-space is an n-tuple of continuous functions $(x_1(t), x_2(t), \ldots, x_n(t))$ of a single variable t.

The graph of a function is a special case; the curve described by the graph of $y = f(x)$ can be written in this "parametric form" by using x for the variable t. Then the curve is $(t, f(t))$. For a given t, we can regard $X(t) = (x_1(t), x_2(t), \ldots, x_n(t))$ either as a point in n-space or as a vector. Thus it is possible to perform vector operations with this quantity. In many of our references to a curve, we shall regard t as time. For instance, we define the velocity at time t of a point on the curve to be

$$X'(t) = (x_1'(t), x_2'(t), \ldots, x_n'(t))$$

if it exists. We define the acceleration at time t to be

$$X''(t) = (x_1''(t), x_2''(t), \ldots, x_n''(t))$$

if it exists.

Since we can really get no adequate intuitive notion of acceleration, we may regard our definition of acceleration as not requiring justification. The definition of velocity is a limiting case of an "average velocity", as follows:

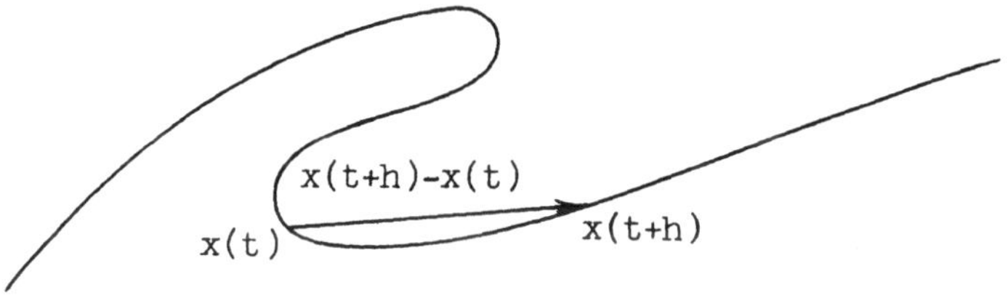

At time t, let the corresponding point of the curve be $X(t)$; at time $t + h$, the point is then $X(t + h)$. If we only know these two points, the simplest type of motion to assume between them is straight-line motion at a constant rate. This means that the vector $X(t + h) - X(t)$ is traversed in time h at a constant rate. This rate, together with its direction, is then given by

$$\frac{X(t+h) - X(t)}{h} = \left(\frac{x_1(t+h) - x_1(t)}{h}, \ldots, \frac{x_n(t+h) - x_n(t)}{h} \right).$$

The velocity at time t is then the limit of this average velocity as the time h over which the average is taken goes to zero, and the result is simply $X'(t)$.

We use the velocity to give us a definition of the tangent to the curve: the tangent line to $X(t)$ at $t = t_0$ is the line through $X(t_0)$ and parallel to the velocity vector at t_0, or $X'(t_0)$. Thus it is the set of points

$$X(t_0) + sX'(t_0),$$

where s ranges through all numbers.

With the concepts we have now developed, we can formulate the basic notions of Newton's mechanics and derive quite easily a number of important consequences. To begin with, we have certain concepts which we shall call primary notions; they are mathematical formulations of quantities observable in nature. They are:

Time (a number t, regarded as measured on a fixed scale).
Position (a vector (or point) function $X(t)$ of time).
Mass (a number $m > 0$ associated with a given point, or "particle").
Force (a vector F).

From these, we derive some further notions, defined as follows:

Velocity: $X'(t)$
Acceleration: $X''(t)$
Momentum: $mX'(t)$
Angular momentum about the point P: $m((X(t) - P) \times X'(t))$ (here and wherever vector products are used, we are treating only 3-space).
Kinetic energy: $\frac{1}{2} m(X'(t))^2$
Moment of the force F acting at the point X about the point P: $(X - P) \times F$.

These notions are subject to certain fundamental laws, which are really those of Newton. The first is simple:

I. $F = mX''(t)$, or in words:
The force acting on a particle at $X(t)$ is the product of the mass of the particle by its acceleration.

The other laws require a little more explanation. A mechanical system is a collection of particles with positions $X_1, X_2, \ldots, X_N$, each depending on time, and having respective masses $m_1, m_2, \ldots, m_N$. Let F_i be the force acting on the i-th particle. Then by I, $F_i = m_i X_i''$. F_i is composed of an external force $\bar{F}_i$, coming from outside the system, and other "internal" forces F_{ij}, where F_{ij} is the force exerted by the j-th particle on the i-th. Thus

$$F_i = \bar{F}_i + F_{i1} + F_{i2} + \cdots + F_{iN} ,$$

which we write more briefly as

$$F_i = \bar{F}_i + \sum_{j=1}^{N} F_{ij} .$$

The other laws concern the F_{ij}:

II. $F_{ij} = - F_{ji}$ ("action equals reaction"; in particular, $F_{ii} = 0$).

III. F_{ij} is parallel to $X_i - X_j$, i.e., the force exerted by one particle on another acts along the line joining them.

Let us consider for a moment the case where all the external forces $\bar{F}_i$ are zero (a <u>closed system</u>). Then

$$m_i X_i'' = F_i = \sum_{j=1}^{N} F_{ij} .$$

If we add up the forces acting on all particles, we see that

$$\sum_{i=1}^{N} m_i X_i'' = \sum_{i=1}^{N} \sum_{j=1}^{N} F_{ij} .$$

Now all $F_{ii} = 0$, and if $i \neq j$, each of F_{ij} and F_{ji} occurs exactly once in the double sum on the right. Since $F_{ij} = - F_{ji}$, their sum is 0, and so the whole double sum is zero.

Thus

$$\sum_{i=1}^{N} m_i X_i'' = 0 .$$

Since $(m_i X_i')' = m_i X_i''$, we have

$$\left(\sum_{i=1}^{N} m_i X_i' \right)' = 0,$$

or

<u>the total momentum</u> $\Sigma m_i X_i'$ <u>is a constant</u> A.

Then $\Sigma m_i X_i = At + B$ for some fixed vector B (all our results for vector-valued functions follow from those for ordinary functions by considering the components separately). If we let $m = \Sigma m_i$ be the total mass and let Y be the vector $\frac{1}{m} \Sigma m_i X_i = \Sigma \frac{m_i}{m} X_i$, we have

$$mY = At + B .$$

The point Y is called the <u>center of mass</u> of the system. We see that <u>in the absence of external forces, the center of mass moves linearly with constant velocity</u>.

If X(t) and Y(t) are vector-valued functions in 3-space, it is left as an exercise to verify that $(X \times Y)' = (X' \times Y) + (X \times Y')$. In particular, $(X \times X')' = (X' \times X') + (X \times X'') = X \times X''$. In a closed system,

$$m_i X_i'' = \sum_{j=1}^{N} F_{ij} ,$$

and so

$$m_i (X_i \times X_i'') = X_i \times m_i X_i'' = \sum_{j=1}^{N} (X_i \times F_{ij}) .$$

Again let us sum over all particles. We shall show that

$$\sum_{i=1}^{N} m_i (X_i \times X_i'') = \sum_{i=1}^{N} \sum_{j=1}^{N} (X_i \times F_{ij}) = 0.$$

As when we showed

$$\sum_{i=1}^{N} \sum_{j=1}^{N} F_{ij} = 0,$$

it will be enough to show that each pair of terms

$$X_i \times F_{ij} \quad \text{and} \quad X_j \times F_{ji}$$

cancel when $i \neq j$. Now $F_{ji} = -F_{ij}$, so the sum of these two is

$$(X_i - X_j) \times F_{ij} .$$

By III, this is 0. Thus our assertion is proved.

But

$$\sum_{i=1}^{N} m_i(X_i \times X_i'') = \left(\sum_{i=1}^{N} m_i(X_i \times X_i') \right)'$$

is the derivative of the total angular momentum (about 0). Hence we conclude that <u>the total angular momentum of a closed system is a constant</u>. (It is left as an exercise in case 0 is replaced by any other point P.)

Now let us consider the case with external forces $\bar{F}_i$. By the fact that the sum of the internal forces is 0, we see that

$$\sum_{i=1}^{N} m_i X_i'' = \sum_{i=1}^{N} \bar{F}_i ,$$

or that

$$\left(\sum_{i=1}^{N} m_i X_i' \right)' = \sum_{i=1}^{N} \bar{F}_i .$$

I.e., <u>the derivative of the total momentum is the sum of the external forces</u>. Since $mY'' = \Sigma m_i X_i''$, we see that the center of mass of the system moves like a particle of mass m acted on by the total external forces.

Also,

$$m_i(X_i \times X_i'') = (X_i \times \bar{F}_i) + \sum_{j=1}^{N} (X_i \times F_{ij}) .$$

Adding as before,

$$\left(\sum_{i=1}^{N} m_i(X_i \times X_i') \right)' = \sum_{i=1}^{N} m_i(X_i \times X_i'') = \sum_{i=1}^{N} (X_i \times \bar{F}_i),$$

or the derivative of the total angular momentum is the sum of the moments of the external forces.

When we have only <u>one</u> particle, all $F_{ij} = 0$, and $\bar{F}_i$ is parallel to X_i. We assume the origin is the source of the force $\bar{F}_i$. We drop the indices i as unnecessary. Then $(m(X \times X'))' = (X \times \bar{F}) = 0$. Applied to the solar system with the assumption that the only non-negligible force acting on a planet is that of the sun, we see that the derivative of the angular momentum of a planet about the sun is 0, or that its angular momentum about the sun is a constant. This is one formulation of Kepler's law that a planet sweeps out area at a constant rate.

15. An Approximation Theorem. Linear Interpolation

Let $f(x)$ be a function having at least n derivatives $f', f'', \ldots, f^{(n)}$ in an interval. Let x_0 be a point of this interval. Then x_0 is called a zero of multiplicity n of $f(x)$ if $f(x_0) = 0$, $f'(x_0) = 0, \ldots, f^{(n-1)}(x_0) = 0$, but $f^{(n)}(x_0) \neq 0$. Thus if $f(x_0) = 0 = f'(x_0) = \ldots = f^{(n-1)}(x_0)$, we can say that x_0 is a zero of $f(x)$ of multiplicity at least n.

If $f(x)$ has zeros (of multiplicity $\geqq 1$) at x_1 and x_2, then by Rolle's Theorem $f'(x)$ has a zero (of multiplicity at least 1) between x_1 and x_2. If $f(x)$ has zeros of multiplicity at least 2 at x_1 and of multiplicity at least 1 at x_2 and, say, $x_1 < x_2$, then $f'(x)$ has a zero of multiplicity at least 1 at x_1 and a zero ξ of multiplicity at least 1 such that $x_1 < \xi < x_2$. Then $f''(x)$ has a zero of multiplicity at least 1 between x_1 and ξ. Likewise, if $f(x)$ has zeros at x_1, x_2, x_3, and $x_1 < x_2 < x_3$, then $f'(x)$ has a zero between x_1 and x_2 and one between x_2 and x_3. Therefore $f''(x)$ has a zero between these zeros of $f'(x)$. Now let us make a general statement.

Let $f(x)$ have n zeros, counting multiplicities, in an interval; we show that $f'(x)$ has at least $(n - 1)$ zeros in the interval, counting multiplicities. If the zeros of $f(x)$ are $x_1 < x_2 < \ldots < x_r$, with respective multiplicities $m_1, m_2, \ldots, m_r$, $m_1 + m_2 + \ldots + m_r = n$, then x_1 is a zero of $f'(x)$ of multiplicity $m_1 - 1$, x_2 a zero of multiplicity $m_2 - 1$, and so on. ("Zeros of multiplicity 0" are simply not zeros.) Moreover, $f'(x)$ has a zero between x_1 and x_2, one between x_2 and x_3, etc., giving at least $r - 1$ more zeros. The total number of zeros, counting multiplicities, of $f'(x)$ in the interval is thus at least

$$(m_1 - 1) + (m_2 - 1) + \ldots + (m_r - 1) + (r - 1)$$

$$= m_1 + m_2 + \ldots + m_r - r + (r - 1) = m_1 + m_2 + \ldots + m_r - 1$$

$$= n - 1.$$

By starting with $f'(x)$ instead of $f(x)$, we see that $f''(x)$ has at least $(n - 2)$ zeros in the interval, $\ldots$, and finally that $f^{(n-1)}(x)$ has at least one zero in the interval. We shall use this result to prove a key theorem which will lead us to several interesting forms of approximations.

THEOREM. Suppose that $f(x)$ has zeros at $x_1, x_2, \ldots, x_r$, of respective multiplicities at least $m_1, m_2, \ldots, m_r$. Let x_0 be such that $f(x)$ has $n = m_1 + m_2 + \ldots + m_r$ derivatives in the smallest interval containing $x_0, x_1, x_2, \ldots, x_r$. Let $H(x) = (x - x_1)^{m_1}(x - x_2)^{m_2} \ldots (x - x_r)^{m_r}$. Then there is a ξ in this interval such that

$$f(x_0) = \frac{H(x_0)}{n!} f^{(n)}(\xi) .$$

PROOF. Suppose first that x_0 is one of $x_1, x_2, \ldots, x_r$. Then $f(x) = 0 = H(x_0)$, and we can take ξ to be any point of the interval. It remains to consider the case where x_0 is different from all of $x_1, x_2, \ldots, x_n$. Then consider

$$g(x) = f(x) - CH(x),$$

where C is a constant such that $g(x_0) = 0$. Note that to be able to choose such a C requires that we can solve for C in $f(x_0) - CH(x_0) = 0$, or simply that $H(x_0) \neq 0$, which we know to be the case. Of course we could write down explicitly what C is, but this is not essential.

Now $g(x)$ has x_1 as a zero of multiplicity at least m_1, x_2 as a zero of multiplicity at least $m_2, \ldots, x_r$ as a zero of multiplicity at least m_r, and x_0 is also a zero of $g(x)$. Thus $g(x)$ has at least $n + 1$ zeros, counting multiplicities, in our interval. Consequently there is a ξ in this interval such that $g^{(n)}(\xi) = 0$. But $g^{(n)}(x) = f^{(n)}(x) - CH^{(n)}(x)$. Observe that when we expand $H(x)$, we get a polynomial with highest term x^n and other terms with lower exponents. But when we take n derivatives, all these other terms become 0 and $(x^n)^{(n)} = n!$. Thus

$$g^{(n)}(x) = f^{(n)}(x) - C \cdot n! ,$$

$$0 = g^{(n)}(\xi) = f^{(n)}(\xi) - C \cdot n! ,$$

and

$$C = \frac{f^{(n)}(\xi)}{n!} .$$

From the fact that $g(x_0) = 0 = f(x_0) - CH(x_0)$, we see that

$$f(x_0) = CH(x_0) = \frac{f^{(n)}(\xi)}{n!} H(x_0),$$

which was to be proved.

As a first application, let us consider the problem of interpolating from tables, for instance from logarithm tables. If we have tables for the function $f(x)$ and we read off $f(x_1)$ and $f(x_2)$ for $x_1 < x_2$, we want to estimate the error introduced by using the straight line through $(x_1, f(x_1))$ and $(x_2, f(x_2))$ in place of the curve for finding $f(x)$ when $x_1 < x < x_2$. Let $y = ax + b$ be the approximating line. Then we consider the function $f(x) - (ax + b)$. This has zeros of multiplicity at least 1 at x_1 and x_2. Therefore we have $n = 2$, and

$$H(x) = (x - x_1)(x - x_2).$$

By our theorem,

$$f(x) - (ax + b) = \frac{(x-x_1)(x-x_2)}{2!} f''(\xi)$$

for some ξ between x_1 and x_2. Now $H(x) \leqq 0$, since $x - x_1$ and $x - x_2$ have opposite signs, and the difference between $f(x)$ and $ax + b$ will be, in absolute value, less than or equal to the absolute value of $H(x)$ at its minimum times the largest possible absolute value for $f''(x)$ in the interval, divided by 2. But $H'(x) = (x - x_1) + (x - x_2)$, which is 0 at the midpoint $x = \frac{1}{2}(x_1 + x_2)$. This is the minimum for $H(x)$, and the value of $H(x)$ there is

$$\frac{(x_2-x_1)(x_1-x_2)}{4} .$$

Thus

$$(*) \qquad |f(x) - (ax + b)| \leqq \frac{(x_2-x_1)^2}{8} \cdot \operatorname{Max} |f''(x)| .$$

For a table of logarithms, with $x_1 = 1.01$, $x_2 = 1.02$, the error is at most

$$\frac{(.01)^2}{8} \cdot \operatorname*{Max}_{1.01 \leqq x \leqq 1.02} \left(\frac{1}{x^2}\right) = \frac{.0001}{8} \frac{1}{1.0201} < .00002 .$$

EXERCISES

1) Give a rule of the form of (*) for extrapolation (where x

is <u>outside</u> the interval $x_1 \leqq x \leqq x_2$).

2) log 70.0 = 4.24850; log 80.0 = 4.38203. Extrapolate to get a value for log 82.0 and estimate your error.

3) log tan 1.00 = 0.1924; log tan 1.10 = 0.2933. Estimate the error in interpolating for any angle between 1.00 and 1.10.

16. Simpson's Rule

We shall now apply our approximation theorem to the computation of definite integrals. This time we shall use a polynomial of third degree to approximate the function we are to integrate. But first let us prove a result about integrals of polynomials:

THEOREM. Let $f(x)$ be a polynomial of degree $\leqq 5$. Let m be the midpoint of the interval $a \leqq x \leqq b$ $(m = \frac{1}{2}(a + b))$. Then

$$\int_a^b f(x)\,dx = \frac{b - a}{6}(f(a) + 4f(m) + f(b)) - \frac{(b - a)^5}{2880} f^{(iv)}(m) .$$

PROOF. Observe first of all that the assertion is <u>linear</u> in f. I.e., if it is true for $f(x)$ and for $g(x)$, then it is true for $f(x) + g(x)$ and for $cf(x)$, where c is a constant. Therefore it will be true for all polynomials of degree $\leqq 5$, if we can show it is true for $1, x, x^2, x^3, x^4, x^5$, or if we can show it for $1, (x - m), (x - m)^2, (x - m)^3, (x - m)^4, (x - m)^5$. For $x^n = ((x - m) + m)^n = (x - m)^n + n(x - m)^{n-1}m + \ldots + n(x - m)m^{n-1} + m^n$ is a polynomial of degree n in $(x - m)$ for any n; thus any polynomial in x of degree $\leqq 5$ is also a polynomial of degree $\leqq 5$ in $(x - m)$.

First let $f(x) = (x - m)^i$, where i is <u>odd</u>: $i = 1, 3, 5$. Then

$$\int_a^b (x - m)^i\,dx = \frac{1}{i + 1}(x - m)\Big|_a^b = \frac{1}{i + 1}\left((b - m)^{i+1} - (a - m)^{i+1}\right)$$

$$= \frac{1}{i + 1}\left(\left(\frac{b - a}{2}\right)^{i+1} - \left(\frac{a - b}{2}\right)^{i+1}\right).$$

But $i + 1$ is <u>even</u>, so

$$\left(\frac{b - a}{2}\right)^{i+1} = \left(\frac{a - b}{2}\right)^{i+1},$$

and

$$\int_a^b (x - m)^i \, dx = 0.$$

On the right side, we have

$$\frac{b - a}{6}\left((a - m)^i + 4(m - m)^i + (b - m)^i\right) - \frac{(b-a)^5}{2880} f^{(iv)}(m) .$$

If $i = 1$ or 3, the fourth derivative of $(x - m)^i$ is 0. If $i = 5$, the fourth derivative is $5!(x - m)$, and $f^{(iv)}(m) = 0$. Thus the right side reduces to

$$\frac{b - a}{6}\left((a - m)^i + (b - m)^i\right) = \frac{b - a}{6}\left(\left(\frac{a - b}{2}\right)^i + \left(\frac{b - a}{2}\right)^i\right) .$$

Since i is <u>odd</u>,

$$\left(\frac{a - b}{2}\right)^i = -\left(\frac{b - a}{2}\right)^i ,$$

and the right-hand side is 0. So in this case, both sides of the equation are 0, and we are done. It remains to treat $f(x) = 1$, $(x - m)^2$ and $(x - m)^4$.

In the first two cases, $f^{(iv)}(x) = 0$ for all x, so we may omit the last term on the right. For $f(x) = 1$,

$$\int_a^b 1 \, dx = b - a,$$

while

$$\frac{b - a}{6}\,(f(a) + 4f(m) + f(b)) = \frac{b - a}{6} \cdot 6 = b - a .$$

Thus the assertion is true for $f(x) = 1$. For $f(x) = (x - m)^2$,

$$\int_a^b f(x)\, dx = \frac{1}{3}\left((b - m)^3 - (a - m)^3\right) = \frac{1}{3}\left(\left(\frac{b - a}{2}\right)^3 - \left(\frac{a - b}{2}\right)^3\right)$$

$$= \frac{2}{3}\left(\frac{b - a}{2}\right)^3 = \frac{(b-a)^3}{12} .$$

Meanwhile,

$$\frac{b-a}{6}(f(a)+4f(m)+f(b)) = \frac{b-a}{6}\left((a-m)^2+(b-m)^2\right)$$

$$= \frac{b-a}{6}\left(\left(\frac{a-b}{2}\right)^2+\left(\frac{b-a}{2}\right)^2\right) = \frac{(b-a)^3}{12} = \int_a^b f(x)\,dx.$$

Finally, let $f(x) = (x-m)^4$. Then $f^{(iv)}(x) = 24$ for all x.

$$\int_a^b f(x)\,dx = \frac{1}{5}\left(\left(\frac{b-a}{2}\right)^5 - \left(\frac{a-b}{2}\right)^5\right) = \frac{1}{80}(b-a)^5 .$$

On the other hand,

$$\frac{b-a}{6}\Big(f(a)+4f(m)+f(b)\Big) - \frac{(b-a)^5}{2880}f^{(iv)}(m)$$

$$= \frac{b-a}{6}\left(\left(\frac{a-b}{2}\right)^4+\left(\frac{b-a}{2}\right)^4\right) - \frac{(b-a)^5}{2880}\cdot 24$$

$$= \frac{(b-a)^5}{48} - \frac{(b-a)^5}{120} = \frac{(b-a)^5}{80} = \int_a^b f(x)\,dx .$$

Thus the assertion is true for all $(x-m)^i$, $i \leqq 5$, and therefore for all polynomials of degree $\leqq 5$.

Now suppose we are given a function $f(x)$ having four derivatives, and suppose we have found a polynomial $P(x)$ of degree 3 such that: $P(a) = f(a)$, $P(b) = f(b)$, $P(m) = f(m)$, $P'(m) = f'(m)$. (We shall show later how $P(x)$ can be constructed. Then $f(x) - P(x)$ has at $x = a$ and at $x = b$ a zero of multiplicity at least one, and at $x = m$ a zero of multiplicity at least 2. Therefore we take the n of the approximation theorem to be 4, and $H(x) = (x-a)(x-m)^2(x-b)$. Then

$$f(x) - P(x) = \frac{H(x)}{4!}\left(f^{(iv)}(\xi) - P^{(iv)}(\xi)\right), \quad a < \xi < b.$$

But $P^{(iv)}(x) = 0$, since $P(x)$ is of degree 3. Now $H(x) \leqq 0$ whenever $a \leqq x \leqq b$, and if Max. denotes the maximum of $f^{(iv)}(x)$ for $a \leqq x \leqq b$, Min. its minimum, we have

$$(**) \qquad \frac{H(x)}{24} \cdot \text{Max.} \leqq f(x) - P(x) \leqq \frac{H(x)}{24} \cdot \text{Min.}$$

Since $H(x)$ is a polynomial of degree $\leqq 5$, and since $H(a) = H(b) = H(m) = 0$, we see by our theorem that

$$\int_a^b H(x)\,dx = \frac{-(b-a)^5}{2880} \cdot H^{(iv)}(m) = \frac{-(b-a)^5}{2880} \cdot 24.$$

Taking the integrals in the inequality (**), we have

$$\frac{-(b-a)^5}{2880} \cdot \text{Max.} \leqq \int f(x)\,dx - \int P(x)\,dx \leqq \frac{-(b-a)^5}{2880} \cdot \text{Min.}$$

If we assume, as we already have in discussing its maximum and minimum, that $f^{(iv)}$ is <u>continuous</u> for $a \leqq x \leqq b$, then by the theorem we quoted in Section 12,

$$\int_a^b f(x)\,dx - \int_a^b P(x)\,dx = \frac{-(b-a)^5}{2880} f^{(iv)}(\xi)$$

for some ξ, $a \leqq \xi \leqq b$. Now since $P^{(iv)}(x) = 0$, we know by our theorem that

$$\int_a^b P(x)\,dx = \frac{b-a}{6}(P(a) + 4P(m) + P(b)) = \frac{b-a}{6}(f(a) + 4f(m) + f(b)).$$

Finally, we have <u>Simpson's rule</u>:

$$\int_a^b f(x)\,dx = \frac{b-a}{6}(f(a) + 4f(m) + f(b)) - \frac{(b-a)^5}{2880} f^{(iv)}(\xi)$$

for some ξ, $a \leqq \xi \leqq b$.

The way in which this rule is used is by using the first term on the right as an approximation to

$$\int_a^b f(x)\,dx.$$

The second term gives an estimate of the error introduced by making this approximation. The error is <u>at most</u>

$$\frac{(b-a)^5}{2880} \cdot |\text{Max}| \quad \text{or} \quad \frac{(b-a)^5}{2880} \cdot |\text{Min}| ,$$

whichever is larger.

Finally, we must show that a polynomial $P(x)$ can always be found satisfying our requirements. Let $Q = (a, f(a))$, $R = (b, f(b))$, $S = (m, f(m))$. Then we can find a straight line $y = Ax + B$ through Q and R (here A and B are simply numbers). Set $g(x) = Ax + B + C(x - a)(x - b)$, where C is so chosen that $g(m) = f(m)$. This is always possible because $(m - a)(m - b) \neq 0$. We still have $g(a) = f(a)$, $g(b) = f(b)$. Now set $P(x) = g(x) + D(x - a)(x - b)(x - m)$. Then $P'(m) = g'(m) + D(m - a)(m - b)$. Since $(m - a)(m - b) \neq 0$, we can choose the constant D so that $P'(m) = f'(m)$. Then $P(a) = g(a) = f(a)$, $P(b) = g(b) = f(b)$, $P(m) = f(m)$ and $P'(m) = f'(m)$; moreover, $P(x)$ is of degree 3. Therefore $P(x)$ is our required polynomial. (Note that in applying Simpson's rule, one need not compute anything about $P(x)$.)

EXERCISES

1) Use Simpson's rule to compute

$$\int_0^1 \frac{dx}{1 + x^2} ;$$

compare with the result you get by direct integration.

2) Compute

$$\int_{-1}^{1} e^{-x^2} dx :$$

a) for $a = -1$, $b = 1$;

b) by using the fact that it is

$$2 \int_0^1 e^{-x^2} dx$$

and using $a = 0$, $b = 1$. Compare your estimates of error in each case.

3) Use Simpson's rule to approximate, with error,

$$\int_{-\frac{1}{2}}^{\frac{1}{2}} \frac{dx}{\sqrt{1 - x^4}} .$$

(No direct means is known of finding indefinite integrals in 2) and 3); therefore approximate methods assume great importance in these cases. 2) is called the <u>error integral</u>, 3) an <u>elliptic integral</u>.)

17. <u>Taylor's Formula</u>.

Suppose now that $f(x)$ has n derivatives in some interval about x_0. Suppose also that we have found a polynomial $P(x)$, of degree $(n-1)$, such that $f(x_0) = P(x_0)$, $f'(x_0) = P'(x_0)$, ..., $f^{(n-1)}(x_0) = P^{(n-1)}(x_0)$. Then $f(x) - P(x)$ has a zero of multiplicity at least n at x_0. By the approximation theorem, if we set $H(x) = (x - x_0)^n$ we have

$$f(x) - P(x) = \frac{H(x)}{n!} f^{(n)}(\xi),$$

or

$$f(x) - P(x) = \frac{(x-x_0)^n}{n!} f^{(n)}(\xi),$$

where ξ is between x and x_0.

It remains to determine $P(x)$. To do so, we write $P(x) = a_0 + a_1(x - x_0) + \cdots + a_{n-1}(x - x_0)^{n-1}$ as a combination of powers of $(x - x_0)$, where the coefficients a_i are yet to be determined. Now we wish to have $P(x_0) = f(x_0)$. But $P(x_0) = a_0$. Therefore $a_0 = f(x_0)$. $P'(x_0) = a_1$, since the derivative of the constant term is 0 and since a factor of $(x - x_0)$ remains in the derivatives of all terms except $a_1(x - x_0)$, which has derivative a_1. Therefore $a_1 = f'(x_0)$. Likewise $P''(x_0) = 2a_2 = f''(x_0)$, and in general $P^{(i)}(x_0) = i!a_i = f^{(i)}(x_0)$, or

$$a_i = \frac{f^{(i)}(x_0)}{i!} .$$

Therefore

$$P(x) = \sum_{i=0}^{n-1} \frac{f^{(i)}(x_0)}{i!} (x - x_0)^i .$$

Summarizing, we have

$$f(x) = \sum_{i=0}^{n-1} \frac{f^{(i)}(x_0)}{i!} (x - x_0)^i + \frac{(x-x_0)^n}{n!} f^{(n)}(\xi) ,$$

where ξ is between x and x_0. This result is known as <u>Taylor's Formula</u>.

We can develop the formula by another approach which will supply some new information of interest. In this case we again assume that $f(x)$

has n derivatives in an interval containing a and b, $a < b$. Then we let

$$F(x) = f(x) + \frac{b-x}{1!} f'(x) + \ldots + \frac{(b-x)^{n-1}}{(n-1)!} f^{(n-1)}(x) .$$

Then

$$F'(x) = f'(x) - f'(x) + \frac{b-x}{1!} f''(x) - \frac{2(b-x)}{2!} f''(x) + \frac{(b-x)^2}{2!} f'''(x)$$

$$- \ldots + \frac{(b-x)^{n-2}}{(n-2)!} f^{(n-1)}(x) - \frac{(n-1)(b-x)^{n-2}}{(n-1)!} f^{(n-1)}(x) + \frac{(b-x)^{n-1}}{(n-1)!} f^{(n)}(x)$$

$$= \frac{(b-x)^{n-1}}{(n-1)!} f^{(n)}(x) ,$$

since the earlier terms all cancel in pairs. Now let $g(x) = f(b) - F(x) - C(b-x)^p$, where $1 \leqq p \leqq n$ and C is chosen so that $g(a) = 0$. Since $F(b) = f(b)$, we also have $g(b) = 0$. By Rolle's Theorem, there is a ξ, $a < \xi < b$, such that

$$g'(\xi) = - F'(\xi) + pC(b-\xi)^{p-1} = 0.$$

Thus

$$C = \frac{F'(\xi)}{p(b-\xi)^{p-1}} = \frac{(b-\xi)^{n-p}}{p(n-1)!} f^{(n)}(\xi),$$

and

$$g(x) = f(b) - F(x) - \frac{(b-x)^p (b-\xi)^{n-p}}{p(n-1)!} f^{(n)}(\xi), \qquad a < \xi < b .$$

In particular, $g(a) = 0$, so

$$f(b) - F(a) = \frac{(b-a)^p (b-\xi)^{n-p}}{p(n-1)!} f^{(n)}(\xi) ,$$

or

$$f(b) = f(a) + \frac{(b-a)}{1!} f'(a) + \ldots + \frac{(b-a)^{n-1}}{(n-1)!} + \frac{(b-a)^p (b-\xi)^{n-p}}{p \cdot (n-1)!} f^{(n)}(\xi),$$

for some ξ between a and b. If we replace a by x_0 and b by x, we see that the remainder we have here is exactly as before if $p = n$. This term is called the <u>Lagrange remainder</u>. For $p = 1$, the remainder term is called the <u>Cauchy remainder</u>. In some cases it is convenient to use one, in some cases the other.

In general, the expression

$$f(x_0) + \frac{(x-x_0)}{1!} f'(x_0) + \frac{(x-x_0)^2}{2!} f''(x_0) + \cdots$$

is called the <u>Taylor's series for</u> $f(x)$ <u>about</u> x_0. When the remainder after n terms, that is, the difference between the first n terms of the series and $f(x)$, approaches 0 as n becomes large, we say that the series <u>converges to</u> $f(x)$. As an example we shall show a series for arctan x about $x_0 = 0$ which converges to arctan x when $-1 \leqq x \leqq 1$. This is really its Taylor's series about 0. We shall use this series to demonstrate a very effective method for computing the decimal expansion of π.

Consider the series

$$\frac{x}{1} - \frac{x^3}{3} + \frac{x^5}{5} + \cdots, \qquad -1 \leqq x \leqq 1.$$

Put

$$f_n(x) = \arctan x - \left(\frac{x}{1} - \frac{x^3}{3} + \cdots \pm \frac{x^{2n-1}}{2n-1} \right).$$

Then

$$f_n'(x) = \frac{1}{1+x^2} - \left(1 - x^2 + x^4 - \cdots \pm x^{2n-2} \right)$$

$$= \frac{1}{1+x^2} - \frac{1-(-x^2)^n}{1+x^2},$$

by the rule for the sum of a geometric series. This is simply

$$\frac{(-x^2)^n}{1+x^2}.$$

When n is <u>even</u>, $f_n'(x) \geqq 0$. In this case $f_n(x)$ is monotone increasing. Since $f_n(0) = 0$, $f_n(x) \geqq 0$ if $x \geqq 0$. Then, observing that when n is even there are an even number of terms in our polynomial, we see that for n even and $x \geqq 0$,

$$\arctan x \geqq \frac{x}{1} - \frac{x^3}{3} + \cdots - \frac{x^{2n-1}}{2n-1}.$$

For n <u>odd</u>, $f_n'(x) \leqq 0$ and $f_n(x) \leqq 0$ for $x \geqq 0$, so

$$\arctan x \leqq \frac{x}{1} - \frac{x^3}{3} + \ldots + \frac{x^{2n-1}}{2n-1} .$$

Since arctan x lies <u>between</u> any two successive polynomials, the difference between a polynomial of k terms and arctan x is at most the difference between this polynomial and the next one, of $(k + 1)$ terms, or

$$\frac{x^{2k+1}}{2k+1} \leqq \frac{1}{2k+1} ,$$

since $0 \leqq x \leqq 1$. Thus for x in this range, <u>the series converges to arctan</u> x. For $x \leqq 0$, we observe that $f_n(-x) = -f_n(x)$; since $f_n(-x)$ approaches zero, so does $f_n(x)$.

For $x = 1$, this gives

$$\frac{\pi}{4} = \arctan 1 = 1 - \frac{1}{3} + \frac{1}{5} - \frac{1}{7} + \ldots .$$

Thus we have a way, but not a very economical one, of computing the decimal expansion of π. However, there is a closely related method which is very effective. Let $a = \arctan \frac{1}{5}$. Then

$$\tan 2a = \frac{2 \tan a}{1 - \tan^2 a} = \frac{\frac{2}{5}}{1 - \frac{1}{25}} = \frac{5}{12} ,$$

and once more,

$$\tan 4a = \frac{\frac{10}{12}}{1 - \frac{25}{144}} = \frac{120}{119} .$$

Let $b = \arctan \frac{1}{239}$. Then

$$\tan (4a - b) = \frac{\tan 4a - \tan b}{1 + \tan 4a \tan b} = \frac{\frac{120}{119} - \frac{1}{239}}{1 + \frac{120}{119} \cdot \frac{1}{239}}$$

$$= \frac{120 \cdot 239 - 119}{119 \cdot 239 + 120} = \frac{119 \cdot 239 + 239 - 119}{119 \cdot 239 + 120} = 1 .$$

Hence $4a - b = \frac{\pi}{4}$, $\pi = 16a - 4b$. Now

$$a = \frac{1}{5} - \frac{1}{3}\left(\frac{1}{5}\right)^3 + \frac{1}{5}\left(\frac{1}{5}\right)^5 - \ldots,$$

and

$$b = \frac{1}{239} - \frac{1}{3}\left(\frac{1}{239}\right)^3 + \frac{1}{5}\left(\frac{1}{239}\right)^5 - \ldots .$$

Using this device, several hundred decimal places of π have been computed.

EXERCISES

1) Let M be any number $\geqq 0$. Then as n becomes large, $\frac{M^n}{n!}$ approaches 0. For if N is an integer, $N \geq 2M$, and $n > N$, then

$$\frac{M^n}{n!} = \left(\frac{M}{1}\frac{M}{2}\cdots\frac{M}{N}\right)\cdot\frac{M}{N+1}\cdot\ \ldots\ \cdot\frac{M}{n}\ .$$

Let

$$\frac{M}{1}\cdot\frac{M}{2}\cdots\frac{M}{N} = A.$$

Then

$$\frac{M}{N+1} < \frac{1}{2},\ \ldots,\ \frac{M}{n} < \frac{1}{2}\ ,$$

and

$$\frac{M^n}{n!} < \frac{A}{2^{n-N}}\ .$$

As n becomes larger,

$$\frac{A}{2^{n-N}} \longrightarrow 0, \quad \text{so} \quad \frac{M^n}{n!} \longrightarrow 0\ .$$

Now show that the series below are the Taylor's series about $x_0 = 0$ for the indicated functions and that they converge to these functions for any x. (Use the Lagrange remainder.)

a) $e^x = 1 + \frac{x}{1!} + \frac{x^2}{2!} + \frac{x^3}{3!} + \cdots$

b) $\sin x = x - \frac{x^3}{3!} + \frac{x^5}{5!} - \frac{x^7}{7!} + \cdots$

c) $\cos x = 1 - \frac{x^2}{2!} + \frac{x^4}{4!} - \frac{x^6}{6!} + \cdots$.

2) Find the Taylor's series about $x_0 = 1$ of $\log x$. For what values of x can you be sure that the series converges to $\log x$?

18. The Bernstein Convergence Theorem. Binomial Series

Let $f(x)$ be a function defined for $a \leqq x \leqq b$; suppose also that for every integer $n \geqq 0$, $f^{(n)}(x)$ exists and $f^{(n)}(x) \geqq 0$ for $a \leqq x \leqq b$. Let x_1 be some point of the interval and choose some fixed positive integer n. Then consider Taylor's formula for $f(x)$ about x_1, where the polynomial of the formula has more than n terms, say m terms. In particular, consider what this gives us for $f(b)$, when we use the Cauchy remainder:

$$f(b) = f(x_1) + \frac{(b-x_1)}{1!} f'(x_1) + \dots + \frac{(b-x_1)^n}{n!} f^{(n)}(x_1) + \dots$$

$$+ \frac{(b-x_1)^{m-1}}{(m-1)!} f^{(m-1)}(x_1) + \frac{(b-x_1)(b-\xi)^{m-1}}{(m-1)!} f^{(m)}(\xi),$$

where $x_1 < \xi < b$. Now all terms of the sum on the right are non-negative; therefore $f(b)$ is greater than or equal to any of them. In particular,

(1) $$\frac{(b-x_1)^n}{n!} f^{(n)}(x_1) \leqq f(b) ,$$

and this is true for any n.

Now consider the function $f''(x)$. Since the derivatives of $f''(x)$ are also derivatives of $f(x)$, they all are non-negative in $a \leqq x \leqq b$. Therefore we can use $f''(x)$ in place of $f(x)$ in (1), obtaining

$$\frac{(b-x_1)^n}{n!} f^{(n+2)}(x_1) \leqq f''(b) , \quad \text{or}$$

(2) $$f^{(n+2)}(x_1) \leqq \frac{n!}{(b-x_1)^n} f''(b) .$$

This last inequality also holds for any n; in particular, it holds for $n - 2$, where n is as in (1). Then it becomes

(3) $$f^{(n)}(x_1) \leqq \frac{(n-2)!}{(b-x_1)^{n-2}} f''(b) .$$

This is true for any point x_1 of the interval.

Now let x_0 be some point of the interval, and consider the Taylor expansion about x_0 for $f(x)$, where x is also in the interval. The Cauchy remainder after n terms has absolute value

$$\frac{|x-x_0| \; |x-\xi|^{n-1}}{(n-1)!} f^{(n)}(\xi) ,$$

where ξ is between x and x_0. By (3), this is less than or equal to

$$\frac{|x-x_0|\ |x-\xi|^{n-1}}{(n-1)!} \cdot \frac{(n-2)!}{(b-\xi)^{n-2}} f''(b)$$

$$= \frac{|x-x_0|\ |x-\xi| \cdot f''(b)}{n-1} \left(\frac{|x-\xi|}{b-\xi}\right)^{n-2} .$$

This will approach 0 as n becomes large if $|x - \xi| \leqq b - \xi$. If $x_0 < x$, then $|x - \xi| = x - \xi \leqq b - \xi$. If $x < x_0$, <u>and if</u> $|x - x_0| \leqq b - x_0$, then $|x - \xi| \leqq |x - x_0| \leqq b - x_0 \leqq b - \xi$. Therefore we know that the remainder approaches zero whenever $|x - x_0| \leqq b - x_0$. Let us summarize the result we have proved, which is the <u>Bernstein convergence theorem</u>.

THEOREM. Let $f(x)$ be such that $f(x) \geqq 0$, $f'(x) \geqq 0$, $f''(x) \geqq 0, \ldots$ for all x such that $a \leqq x \leqq b$. Let $a < x_0 < b$, and let x be a point of the interval such that $|x - x_0| \leqq b - x_0$. Then the Taylor's series

$$f(x_0) + \frac{(x-x_0)}{1!} f'(x_0) + \frac{(x-x_0)^2}{2!} f''(x_0) + \cdots$$

converges to $f(x)$.

For instance, this theorem tells us that the Taylor's series for e^x formed about any point at all converges to e^x for all x. The result can be extended to a considerably larger class of functions if we make a few observations. First suppose there is some number M such that $f(x) \geqq M$, $f'(x) \geqq M$, $f''(x) \geqq M, \ldots$ whenever $a \leqq x \leqq b$. If $M \geqq 0$, then the Taylor's series of $f(x)$ converges to $f(x)$ by the Bernstein theorem. If $M < 0$, let

$$g(x) = f(x) - Me^{(x-a)} .$$

Then $g(x) \geqq M - Me^{x-a} = M(1 - e^{x-a}) \geqq 0$, since both factors are less than or equal to zero. Likewise, $g'(x) = f'(x) - Me^{x-a} \geqq 0$, $g''(x) \geqq 0, \ldots$. Therefore the Taylor's series for $g(x)$ converges to $g(x)$. As we have remarked, the Taylor's series for e^{x-a} converges to $e^{x-a} (= e^{-a} \cdot e^x)$. Since the operation of forming the Taylor's series is a <u>linear</u> one, the Taylor's series for $f(x)$ is that for $g(x)$ plus M times that for e^{x-a}; the remainder in each of these goes to zero, and therefore the remainder for $f(x)$, which is the sum of these remainders, goes to zero. Thus we see that the Taylor's series for $f(x)$ converges to $f(x)$ if all derivatives of $f(x)$ are <u>bounded from below</u>.

Next suppose $f^{(n)}(x) \geqq 0$ for $a \leqq x \leqq b$, whenever $n > N$ for some fixed N. Then let M be the smallest of the numbers

$$0, \text{Min. } f(x), \text{Min. } f'(x), \ldots, \text{Min. } f^{(N)}(x) \quad \text{on} \quad a \leqq x \leqq b .$$

Then $f^{(i)}(x) \geqq M$ for all $i \geqq 0$, and we see as above that the Taylor's series for $f(x)$ converges to $f(x)$. Since the Taylor's series for $-f(x)$ has as terms the negatives of those in the Taylor's series for $f(x)$, we can use $-f(x)$ to prove that the Taylor's series for $f(x)$ converges to $f(x)$ whenever all $f^{(n)}(x)$ are <u>bounded from above</u>, in particular if all $f^{(n)}(x) \leqq 0$.

Now let $f(x) \geqq 0$, $f'(x) \leqq 0$, $f''(x) \geqq 0$, $f'''(x) \leqq 0, \ldots$; in other words, the derivatives of $f(x)$ <u>alternate in sign</u>. Let $a < x_0 < b$. Then $-b < -x_0 < -a$, and if $-b \leqq \xi \leqq -a$, let $g(\xi) = f(-\xi)$. Then

$$g(\xi) \geqq 0, \; g'(\xi) = -f'(-\xi) \geqq 0, \; g''(\xi) \geqq 0, \; \ldots ,$$

or $g^{(n)}(\xi) \geqq 0$ for all n whenever $-b \leqq \xi \leqq -a$. Thus we know by the Bernstein theorem that the Taylor's series for $g(\xi)$ about $-x_0$ converges to $g(\xi)$ whenever $|\xi - (-x_0)| \leqq -a - (-x_0)$, or whenever $|\xi + x_0| \leqq x_0 - a$. The Taylor's series for $g(\xi) = f(-\xi)$ about $-x_0$ is:

$$g(-x_0) + \frac{(\xi+x_0)}{1!} g'(-x_0) + \frac{(\xi+x_0)^2}{2!} g''(-x_0) + \ldots$$

$$= f(x_0) - \frac{(\xi+x_0)}{1!} f'(x_0) + \frac{(\xi+x_0)^2}{2!} f''(x_0) - \ldots .$$

Now let $x = -\xi$, $\xi = -x$. Then whenever $|-x + x_0| \leqq x_0 - a$, or $|x - x_0| \leqq x_0 - a$, we have, in the limit,

$$f(x) = g(\xi) = f(x_0) - \frac{(-x+x_0)}{1!} f'(x_0) + \frac{(-x+x_0)^2}{2!} f''(x_0) - \ldots$$

$$= f(x_0) + \frac{(x-x_0)}{1!} f'(x_0) + \frac{(x-x_0)^2}{2!} f''(x_0) + \ldots .$$

But this is exactly the Taylor's series for $f(x)$. Note that in this case it converges to $f(x)$ whenever x is closer to x_0 than x_0 is to a, while in the original theorem we had convergence to $f(x)$ whenever x was closer to x_0 than x_0 was to b. Observe also that as before it is enough that the signs alternate from some N on.

We shall apply this last observation to investigate the function $f(x) = x^\alpha$ for $x > 0$, α any number. Then

$$f'(x) = \alpha x^{\alpha-1}, \quad f''(x) = \alpha(\alpha - 1)x^{\alpha-2}, \quad \ldots,$$

and eventually

$$f^{(N)}(x) = \alpha(\alpha - 1) \ldots (\alpha - N + 1)x^{\alpha-N}, \quad \text{where} \quad \alpha < N.$$

Then $f^{(N+1)}(x) = \alpha(\alpha - 1) \ldots (\alpha - N + 1)(\alpha - N)x^{\alpha-N-1}$ has the opposite sign from $f^{(N)}(x)$ for all $x > 0$, since $\alpha - N < 0$, $f^{(N+2)}(x)$ has the opposite sign from $f^{(N+1)}(x)$, and the signs of $f^{(n)}(x)$ alternate for $n > N$. Therefore if ϵ is any small positive number and $x_0 > \epsilon$, the Taylor's series for $f(x)$ about x_0 converges to $f(x)$ provided $|x - x_0| \leqq x_0 - \epsilon$. If $|x - x_0| < x_0$, then we are sure that $|x - x_0| \leqq x_0 - \epsilon$ for some positive ϵ. Therefore the series converges to $f(x)$ provided $|x - x_0| < x_0$.

Now let us change notation a little and write $x_0 = a$, $x = a + b$. Then we expand $f(x) = (a + b)^\alpha$ in powers of $(x - x_0) = b$. (Here $a > 0$.) Thus

$$x^\alpha = (a + b)^\alpha = a^\alpha + \alpha a^{\alpha-1}b + \frac{\alpha(\alpha-1)}{2} a^{\alpha-2}b^2 + \ldots$$

$$+ \frac{\alpha(\alpha-1) \ldots (\alpha-n)}{n!} a^{\alpha-n}b^n + \ldots$$

in the sense that the series converges to $(a + b)^\alpha$ whenever $|(a + b) - a| < a$, or whenever $|b| < a$. This is the so-called <u>binomial series</u>, and the coefficients

$$\frac{\alpha(\alpha-1) \ldots (\alpha-n)}{n!} = \binom{\alpha}{n}$$

are called the <u>binomial coefficients</u>. Observe that if α is a positive integer, all the coefficients are 0 from some point on, and we have the ordinary binomial theorem.

NOTE: In early investigations of Taylor's series, people often simply assumed that the series approached the function whenever the series approached any kind of limit. That this is false is seen by considering the function

$$f(x) = \begin{Bmatrix} e^{-\frac{1}{x^2}}, & x \neq 0 \\ 0, & x = 0 \end{Bmatrix}.$$

Then $f(0) = 0 = f'(0) = f''(0) = \ldots = 0$, so the Taylor's series about $x_0 = 0$ is simply a series of 0's, having limit 0. However, the function is never 0 except at $x = 0$.

EXERCISES

1) Apply the Bernstein theorem to show that the Taylor's series for $\sin x$ and $\cos x$ about any x_0 converge to the respective functions for all x.

2) Apply the theorem to show that the Taylor's series for $\log x$ about $x_0 = 1$ converges to $\log x$ if $|x - 1| < 1$.

3) Use $a = 32$, $b = 1$ in the binomial series to estimate $\sqrt[5]{33}$ to four decimal places.

4) Use $a = 100$, $b = -2$ to estimate $\sqrt{98}$. Since this is also $7\sqrt{2}$, obtain an estimate of $\sqrt{2}$ to five places.

19. Functions of Several Variables

By a function $f(x_1, x_2, \ldots, x_n)$ of n variables, we mean of course a rule which assigns to each point $(x_1^{(0)}, x_2^{(0)}, \ldots, x_n^{(0)})$ of some region in n-space a single number, which we denote by $f(x_1^{(0)}, x_2^{(0)}, \ldots, x_n^{(0)})$. Let us for the moment consider functions of two variables $f(x, y)$. It would be possible to represent this function graphically in three dimensions, by plotting the set of points $(x, y, f(x, y))$ for (x, y) in the domain of the function. But since attempts to represent 3-space on the blackboard are awkward at best, we choose instead to draw "contour map" for the function; that is, we plot on the blackboard the set of points (x, y) where $f(x, y) = 0$, the set where $f(x, y) = 7$, or in general, the set of points (x, y) where $f(x, y) = a$, where a is any number in the range of $f(x, y)$. In general, these sets of points will be curves, and are customarily called the level curves of $f(x, y)$, or less precisely, the level lines of the contour map. These curves represent paths which lie at constant height on the surface representing the function in 3-space.

Let us consider the contour map for the function

$$f(x, y) = \begin{cases} \dfrac{xy^2}{x^2 + y^4}, & (x, y) \neq (0, 0) \\ 0, & (x, y) = (0, 0) \end{cases}.$$

Consider the parabola $x = \alpha y^2$. For (x, y) on this curve, $\alpha \neq 0$, $y \neq 0$, we have

$$f(x, y) = \frac{\alpha y^4}{(\alpha^2+1)y^4} = \frac{\alpha}{\alpha^2 + 1},$$

or $f(x, y)$ is a constant. By solving for x in terms of y in the equation

$$\frac{xy^2}{x^2 + y^4} = \beta, \qquad \beta \neq 0,$$

we see that each level curve of height $\beta \neq 0$, is just a parabola $x = \alpha y^2$ with the point $(0, 0)$ omitted. Thus

$$\beta = \frac{\alpha}{\alpha^2 + 1}.$$

Now what numbers can have the form

$$\frac{\alpha}{\alpha^2 + 1} \quad ?$$

To answer this question, we investigate the function

$$g(\alpha) = \frac{\alpha}{\alpha^2 + 1}$$

for maxima and minima, finding that it has a maximum when $\alpha = 1$, a minimum when $\alpha = -1$. Thus

$$-\frac{1}{2} \leqq \frac{\alpha}{\alpha^2 + 1} \leqq \frac{1}{2}.$$

The level curve of minimum height $(-\frac{1}{2})$ is the parabola $y = -x^2$, that of maximum height $(+\frac{1}{2})$ is the parabola $y = x^2$ (always omitting $(0, 0)$). Thus the contour map looks like the one on the following page.

Observe that for any fixed value of x, this function $f(x, y)$ is continuous in y, and that it is also continuous in x when y is held fixed. It is also continuous along any line; that is, if (x_0, y_0) is a point on a line, then for (x, y) near to (x_0, y_0) <u>and on the line,</u> $f(x, y)$ is near $f(x_0, y_0)$. However, we can <u>not</u> call this function continuous at $(0, 0)$. For that would mean that whenever (x, y) was near to $(0, 0)$, $f(x, y)$ would be near $f(0, 0) = 0$. But there are points on the parabola $y = x^2$ as near as we wish to $(0, 0)$, and at these $f(x, y) = \frac{1}{2}$. The surface has a ridge along $y = x^2$, $x \neq 0$; a trough along

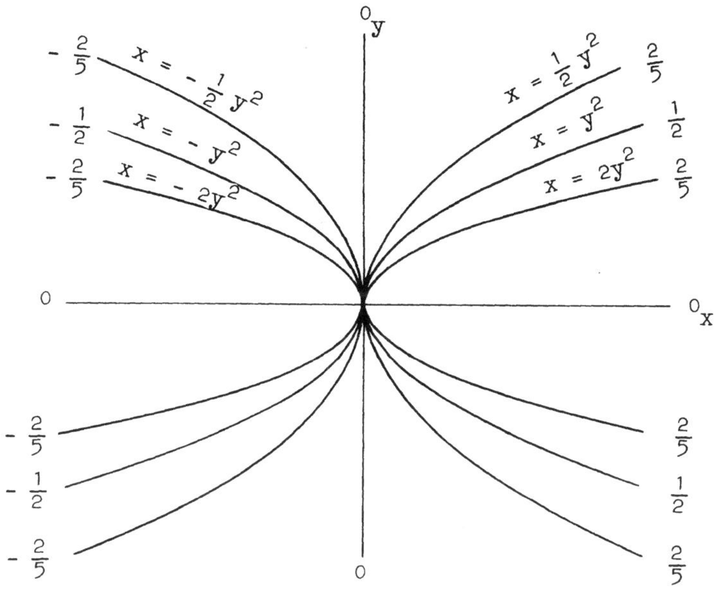

$y = -x^2$, $x \neq 0$; a shallow trough along $y = 0$ for $x > 0$; and a mild ridge along $y = 0$ for $x < 0$. It has a <u>discontinuity</u> at $(0, 0)$.

Let us make the definition of continuity more precise for these functions so that we can discuss it intelligently:

> $f(x, y)$ is called <u>continuous</u> at (x_0, y_0) if for <u>any</u> positive number ϵ, we can find a <u>neighborhood</u> of (x_0, y_0) such that whenever (x, y) is in this neighborhood, then
>
> $$|f(x, y) - f(x_0, y_0)| < \epsilon .$$

A neighborhood may be regarded as a circular disk with (x_0, y_0) as center or as a square with (x_0, y_0) as center. If the function discussed above were to be continuous at $(0, 0)$, then presumably we could start with $\epsilon = \frac{1}{4}$ and find a circle about $(0, 0)$ such that for all points (x, y) inside it,

$$|f(x, y) - 0| < \frac{1}{4} .$$

But inside every such circle we can find points (other than $(0, 0)$) of the parabola $x = y^2$, where $f(x, y) = \frac{1}{2}$. Then we should have to have

$$\frac{1}{2} = \left|\frac{1}{2} - 0\right| < \frac{1}{4} ,$$

which is nonsense.

As a second example, about which we shall say more a little later, let

$$f(x, y) = \begin{cases} \dfrac{4xy(x^2-y^2)}{x^2 + y^2} , & (x, y) \neq (0, 0) \\ 0 , & (x, y) = (0, 0) \end{cases} .$$

(In terms of <u>polar coordinates</u> (r, θ), where $x = r \cos \theta$, $y = r \sin \theta$, $f(x, y) = r^2 \sin 4\theta$.) Though we shall not describe the level curves in detail, the contour map is roughly as follows:

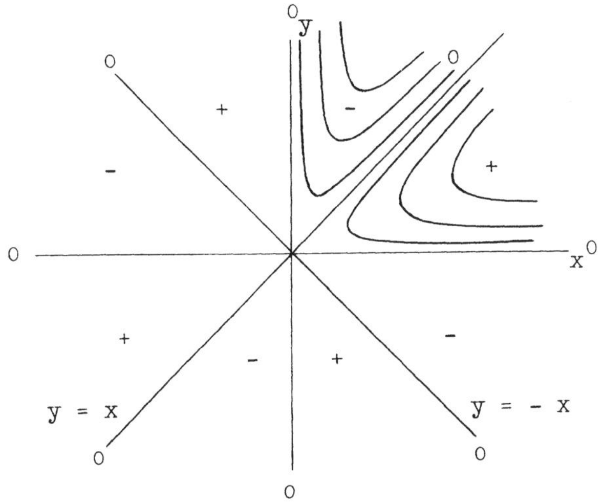

The + signs in the contour map indicate sectors like the one with a + sign in which the contour lines are sketched, and similarly for the - signs. Thus if we stand at $(0, 0)$, we see four valleys sloping away, alternating with four mountains. This function is <u>continuous everywhere</u>.

EXERCISES

Draw a contour map and discuss continuity for each of the following functions:

a) $f(x, y) = xy$,

b) $$f(x, y) = \begin{cases} \dfrac{2x(x^2+y^2)}{x^2 + (x^2+y^2)^2}, & (x, y) \neq (0, 0) \\ 0, & (x, y) = (0, 0) \end{cases}.$$

20. Partial Derivatives

Now let $f(x, y)$ be a function of two variables. We consider the problem of finding derivatives of $f(x, y)$. For the present, we shall consider two kinds of derivatives: if (x_0, y_0) is a point, we shall call the derivative at x_0 of the function $g(x) = f(x, y_0)$ the partial derivative of $f(x, y)$ with respect to x at (x_0, y_0) and write it as

$$f_x(x_0, y_0), \quad \text{or} \quad \frac{\partial f}{\partial x}(x_0, y_0).$$

Likewise, the partial derivative of $f(x, y)$ with respect to y at (x_0, y_0), written as

$$f_y(x_0, y_0), \quad \text{or} \quad \frac{\partial f}{\partial y}(x_0, y_0)$$

is the derivative of the function $h(y) = f(x_0, y)$ at $y = y_0$. Thus we have

$$f_x(x_0, y_0) = \lim_{h \to 0} \frac{f(x_0+h, y_0) - f(x_0, y_0)}{h},$$

$$f_y(x_0, y_0) = \lim_{k \to 0} \frac{f(x_0, y_0+k) - f(x_0, y_0)}{k}.$$

Then it should be evident what we mean by f_{xx}, f_{xy}, f_{yx}, f_{yy}; for instance, $f_{xy}(x_0, y_0) = (f_x)_y(x_0, y_0)$, or in words, the partial derivative with respect to y of the partial derivative with respect to x of $f(x, y)$. To find it, we must use

$$f_{xy}(x_0, y_0) = \lim_{k \to 0} \frac{f_x(x_0, y_0+k) - f_x(x_0, y_0)}{k}$$

$$= \lim_{k \to 0} \lim_{h \to 0} \frac{\dfrac{f(x_0+h, y_0+k) - f(x_0, y_0+k)}{h} - \dfrac{f(x_0+h, y_0) - f(x_0, y_0)}{h}}{k}$$

$$= \lim_{k \to 0} \lim_{h \to 0} \frac{f(x_0+h,y_0+k) - f(x_0,y_0+k) - f(x_0+h,y_0) + f(x_0,y_0)}{hk}$$

Our numerator is thus a combination of the values of the function at the four corners of the rectangle shown below, with the corresponding signs:

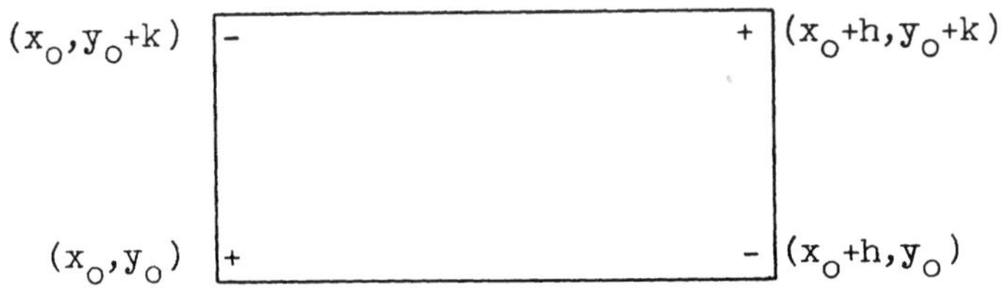

Likewise,

$$f_{yx}(x_0,y_0) = \lim_{h \to 0} \lim_{k \to 0} \frac{f(x_0+h,y_0+k) - f(x_0+h,y_0) - f(x_0,y_0+k) + f(x_0,y_0)}{hk}$$

The fact that the two terms which are involved in taking limits are equal does <u>not</u> mean that the limits are equal; for <u>we are not taking the same limit in both cases</u>. For example, consider

$$\lim_{h \to 0} \lim_{k \to 0} \frac{h^2 - k^2}{h^2 + k^2} = \lim_{h \to 0} \frac{h^2}{h^2} = \lim_{h \to 0} 1 = 1.$$

On the other hand,

$$\lim_{k \to 0} \lim_{h \to 0} \frac{h^2 - k^2}{h^2 + k^2} = \lim_{k \to 0} \frac{-k^2}{k^2} = \lim_{k \to 0} -1 = -1.$$

We do know, by the existence of f_x, that we can apply the mean value theorem to $f(x, y)$ when considered as a function of x for finding y. Thus we have

$$f(x_0 + h, y_0 + k) - f(x_0, y_0 + k) = hf_x(\xi, y_0 + k),$$

where ξ is between x_0 and $x_0 + h$, and

$$f(x_0 + h, y_0) - f(x_0, y_0) = hf_x(\xi', y_0),$$

where ξ' is between x_0 and $x_0 + h$. Suppose for the moment that $\xi = \xi'$; then the numerator in the expressions whose limits are the mixed second partial derivatives is

$$h\Big(f_x(\xi, y_0 + k) - f_x(\xi, y_0)\Big) = hkf_{xy}(\xi, \eta)$$

for some η between y_0 and $y_0 + k$. Thus we would have, for instance,

$$f_{yx}(x_0, y_0) = \lim_{h \to 0} \lim_{k \to 0} f_{xy}(\xi, \eta),$$

where (ξ, η) is in our rectangle. Then if we knew that

$$\lim_{h \to 0} \lim_{k \to 0} f_{xy}(\xi, \eta) = f_{xy}(x_0, y_0),$$

we would have the result that

$$f_{xy}(x_0, y_0) = f_{yx}(x_0, y_0) .$$

However, this is <u>not always true</u>. It is left as an exercise to show that for the function

$$f(x, y) = \left\{ \begin{array}{ll} \dfrac{4xy(x^2-y^2)}{x^2 + y^2}, & (x, y) \neq (0, 0) \\ \\ 0 \quad , & (x, y) = (0, 0) \end{array} \right\},$$

we have $f_{xy}(0, 0) = +4$, $f_{yx}(0, 0) = -4$. But in this example, neither f_{xy} nor f_{yx} is <u>continuous</u> at $(0, 0)$. If f_{xy} <u>is</u> continuous at (x_0, y_0). then we do have

$$\lim_{h \to 0} \lim_{k \to 0} f_{xy}(\xi, \eta) = f_{xy}(x_0, y_0),$$

and we would have shown $f_{xy}(x_0, y_0) = f_{yx}(x_0, y_0)$ if we could be sure that $\xi = \xi'$. But we can arrange this:

Let

$$g(x) = f(x, y_0 + k) - f(x, y_0) .$$

Then

$$g(x_o + h) - g(x_o) = f(x_o + h, y_o + k) - f(x_o + h, y_o)$$
$$- f(x_o, y_o + k) + f(x_o, y_o)$$

is exactly the numerator of the expression we are considering. Since $g(x)$ is differentiable, with $g'(x) = f_x(x, y_o + k) - f_x(x, y_o)$, we have

$$g(x_o + h) - g(x_o) = hg'(\xi),$$

where ξ is between x_o and $x_o + h$. Thus

$$f(x_o + h, y_o + k) - f(x_o + h, y_o) - f(x_o, y_o + k) + f(x_o, y_o)$$
$$= h\Big(f_x(\xi, y_o + k) - f_x(\xi, y_o)\Big).$$

Now this is equal to $hkf_{xy}(\xi, \eta)$ for some η between y_o and $y_o + k$, and we can finish the proof as when we assumed $\xi = \xi'$. What we have proved is the following: <u>Let</u> $f_{xy}(x_o, y_o)$ <u>and</u> $f_{yx}(x_o, y_o)$ <u>exist, and let either</u> f_{xy} <u>or</u> f_{yx} <u>be continuous at</u> (x_o, y_o). <u>Then</u>

$$f_{yx}(x_o, y_o) = f_{xy}(x_o, y_o) .$$

In practically all applications, we do have $f_{yx} = f_{xy}$, so you will see this property used freely in most books.

EXERCISES

Courant, p. 472, ex. 1, 2.

21. Total Differential and Gradient

Let $f(x, y)$ be a function of two variables. Suppose the first variable x is changed by an amount dx and the second by an amount dy. The value of the function at the new point is then $f(x + dx, y + dy)$. The amount that the value of the function has changed is

$$f(x + dx, y + dy) - f(x, y).$$

This expression is a function of the <u>four</u> variables x, y, dx, dy. It is called the <u>total differential</u> of $f(x, y)$, and is written simply as df:

$$df = f(x + dx, y + dy) - f(x, y).$$

Often the total differential is discussed in terms of "infinitesimals", this term referring to dx and dy. However, no adequate definition of an infinitesimal is ever given. The only way we can work with dx and dy is to let these quantities be <u>numbers</u>, which is what we do. It is also commonly written that

$$df = f_x dx + f_y dy ,$$

where f_x and f_y refer to $f_x(x, y)$ and $f_y(x, y)$. We shall see that this is really only an <u>approximate</u> formula, and we shall obtain some idea of the error in its application. There is no objection to the use of the formula, provided one interprets it in this sense.

Now let us reconsider the total differential. To begin with, we write $df = f(x + dx, y + dy) - f(x, y)$ as

$$f(x + dx, y + dy) - f(x + dx, y) + f(x + dx, y) - f(x, y).$$

Then we apply the mean value theorem to get

$$df = f_y(x + dx, \eta)\, dy + f_x(\xi, y)\, dx,$$

where ξ is between x and $x + dx$ and η is between y and $y + dy$. We assume that f_x and f_y are continuous. Then

$$f_y(x + dx, \eta) - f_y(x, y) = \alpha(dx, dy),$$

where $\alpha(dx, dy)$ goes to zero as $\sqrt{dx^2 + dy^2}$ goes to zero (since $(x + dx, \eta)$ is in the rectangle with vertices (x, y), $(x + dx, y)$, $(x + dx, y + dy)$, $(x, y + dy)$, and so approaches (x, y) as $\sqrt{dx^2 + dy^2}$, the diagonal of the rectangle, approaches 0). Likewise,

$$f_x(\xi, y) - f_x(x, y) = \beta(dx, dy)$$

approaches 0 as $\sqrt{dx^2 + dy^2}$ approaches 0. Our formula now becomes

$$df = f_x(x, y)\, dx + f_y(x, y)\, dy + \beta(dx, dy)\, dx + \alpha(dx, dy)\, dy.$$

Now if $\epsilon > 0$ is any small positive number, we can choose dx and dy so that $\sqrt{dx^2 + dy^2}$ is so small that $|\alpha(dx, dy)| < \epsilon$ and $|\beta(dx, dy)| < \epsilon$, for this is just the meaning of the statement that these quantities go to zero as $\sqrt{dx^2 + dy^2}$ goes to zero. Then we have

$$|\alpha(dx, dy)\, dy + \beta(dx, dy)\, dx| \leqq |\alpha(dx, dy)|\ |dy|$$

$$+ |\beta(dx, dy)|\ |dx| < \epsilon|dy| + \epsilon|dx|$$

$$= \epsilon(|dx| + |dy|).$$

Now $|dx| + |dy| \leqq 2\sqrt{dx^2 + dy^2}$, so that when dx and dy are chosen small enough, the error in using $f_x(x, y)\, dx + f_y(x, y)\, dy$ to approximate the total differential is such that

$$\frac{\text{error}}{\sqrt{dx^2 + dy^2}} < \frac{2\epsilon\sqrt{dx^2 + dy^2}}{\sqrt{dx^2 + dy^2}} = 2\epsilon .$$

In other words,

$$\frac{\text{error}}{\sqrt{dx^2 + dy^2}} \longrightarrow 0 \quad \text{as} \quad \sqrt{dx^2 + dy^2} \longrightarrow 0 .$$

This is exactly what we shall mean when we say, "The error is small when compared to $\sqrt{dx^2 + dy^2}$". Thus our result is the following:

$$df = f_x(x, y)\, dx + f_y(x, y)\, dy ,$$

with an error which is small when compared to $\sqrt{dx^2 + dy^2}$. This is then the sense in which the formula is correctly interpreted.

The expression $f_x dx + f_y dy$ which occurs in the above formula can be expressed in vector notation as the scalar product

$$(f_x, f_y) \cdot (dx, dy) .$$

There is little reason to do this in two dimensions. However, we can define the total differential of a function $f(x_1, x_2, \ldots, x_n)$ of n variables as the quantity

$$f(x_1 + dx_1, x_2 + dx_2, \ldots, x_n + dx_n) - f(x_1, x_2, \ldots, x_n),$$

a function of $2n$ variables, and prove as above that

$$df = (f_{x_1}, f_{x_2}, \ldots, f_{x_n}) \cdot (dx_1, dx_2, \ldots, dx_n) + \text{error}$$

where the error is small compared to

$$\sqrt{dx_1^2 + dx_2^2 + \dots + dx_n^2} \ .$$

We denote the vector $(dx_1, dx_2, \dots, dx_n)$ by dX, and we denote the vector $(f_{x_1}, f_{x_2}, \dots, f_{x_n})$ by <u>grad f</u> (called "the <u>gradient</u> of f"). In this notation, we have

$$df = (\text{grad } f) \cdot dX + \text{error},$$

where the error is small compared to $|dX|$.

Now, for a change, let $f(x, y, z)$ be a function of three variables, and let each of these be a function $x(t), y(t), z(t)$ of a single variable t. Then $g(t) = f(x(t), y(t), z(t))$ is a function of t, and we can investigate $g'(t)$. To do so, we must consider

$$\frac{g(t+h) - g(t)}{h} = \frac{f(x(t+h), y(t+h), z(t+h)) - f(x(t), y(t), z(t))}{h} \ .$$

Now let $dx = x(t + h) - x(t)$, $dy = y(t + h) - y(t)$, $dx = z(t + h) - z(t)$; then we have

$$\frac{g(t+h) - g(t)}{h} = \frac{df}{h} = \frac{(\text{grad } f) \cdot dX}{h} + \frac{\text{error}}{h} \ .$$

Now

$$\frac{1}{h} dX = \left(\frac{x(t+h) - x(t)}{h}, \frac{y(t+h) - y(t)}{h}, \frac{z(t+h) - z(t)}{h} \right),$$

so that

$$\lim_{h \to 0} (\text{grad } f) \cdot \frac{dX}{h} = (\text{grad } f) \cdot X'(t) \ .$$

The error term is of the form $\alpha dx + \beta dy + \gamma dz$, where α, β and γ go to zero as $|dX|$ goes to zero. As $h \to 0$, $|dX| \to 0$, since $x(t), y(t), z(t)$ are assumed continuous. Thus

$$\frac{\text{error}}{h} = \underset{\substack{\downarrow \\ 0}}{\alpha} \underset{\substack{\downarrow \\ x'(t)}}{\frac{dx}{h}} + \underset{\substack{\downarrow \\ 0}}{\beta} \underset{\substack{\downarrow \\ y'(t)}}{\frac{dy}{h}} + \underset{\substack{\downarrow \\ 0}}{\gamma} \underset{\substack{\downarrow \\ z'(t)}}{\frac{dz}{h}}$$

and so

$$\frac{\text{error}}{h} \to 0 \quad \text{as} \quad h \to 0,$$

as indicated by the vertical arrows.

Substituting for $g(t)$, we have the formula

$$\frac{d}{dt}(f(x(t), y(t), z(t))) = (\text{grad } f) \cdot X'(t) \quad ,$$

which is the 3-variable case of the chain rule for functions of several variables. (In the case of $f(x(t))$, it becomes simply $f'(x) \cdot x'(t)$, as we have seen before.)

EXERCISE

Let $f(x, y)$ be a function of two variables. Suppose there exists a positive number M such that in some neighborhood of (x_o, y_o), f_x and f_y exist and $|f_x(x, y)| < M$, $|f_y(x, y)| < M$ for all (x, y). By considering the total differential, prove that $f(x, y)$ is continuous at (x_o, y_o). Then use this result to prove that if f_x and f_y are continuous at (x_o, y_o), $f(x, y)$ is also continuous there.

22. Applications

A. Errors. First of all, we mention the use of the total differential to estimate the error in a quantity computed from quantities in which the error is known. For instance, the gravitational attraction between masses m_1, m_2 at distance r is given by

$$\frac{m_1 m_2}{r^2} = f(m_1, m_2, r).$$

We might ask: If we have found m_1 and m_2 each to be 1000 grams $\pm$ .01 g. and $r = 10$ cm. $\pm$ .001 cm., then how large an error can result in calculating the force, using these numbers? Since we are only interested in an estimate, we neglect the error term in the expression for df and find

$$df = \frac{m_2}{r^2}\, dm_1 + \frac{m_1}{r^2}\, dm_2 - \frac{2m_1 m_2}{r^3}\, dr$$

$$= \frac{10^3}{10^2}\, dm_1 + \frac{10^3}{10^2}\, dm_2 - \frac{2 \cdot 10^6}{10^3}\, dr \ .$$

The absolute value of this is greatest when $dm_1 = + .01 = dm_2$, $dr = - .001$. Then we have

$$df = 10 \cdot (.01) + 10 \cdot (.01) + 2 \cdot 10^3 \cdot (.001) = 2.2 \ .$$

Thus our estimate of the maximum error in the force is $\pm$ 2.2 dynes.

B. Maxima and minima. We can make a simple test for relative maxima and minima of functions of several variables. We shall speak in terms of functions of two variables, but the extension to functions of more variables will be clear. A point (x_o, y_o) is called a relative maximum of $f(x, y)$ if there is a neighborhood of (x_o, y_o) such that for all points (x, y) of this neighborhood,

$$f(x, y) \leqq f(x_o, y_o).$$

A relative minimum is similarly defined. Assume that f_x and f_y exist at the relative maximum (x_o, y_o). Then the function $g(x) = f(x, y_o)$ has a relative maximum at x_o, so $g'(x_o) = f_x(x_o, y_o) = 0$. Likewise, $h(y) = f(x_o, y)$ has a relative maximum at y_o, so $h'(y_o) = f_y(x_o, y_o) = 0$. Thus we conclude:

If $f(x, y)$ has a relative maximum or minimum at (x_o, y_o), then

$$f_x(x_o, y_o) = 0 = f_y(x_o, y_o) .$$

The converse is no more true than in the case of functions of one variable. There are higher derivative tests as before, but we shall not discuss them here.

C. Tangent planes and normal lines to surfaces. Next consider the set of points (x, y, z) such that $f(x, y, z) = c$, where c is some constant and $f(x, y, z)$ has continuous first derivatives. Such a set of points is called a level surface for the function $f(x, y, z)$. It is completely analogous to the level curves for the functions $f(x, y)$ of two variables which we used in drawing our contour maps. A particular case of a level surface is the "graph" of a function $g(x, y)$ of two variables. The points we plot in this case are the points $(x, y, g(x, y))$, and these are exactly the level surface $f(x, y, z) = 0$ where

$$f(x, y, z) = z - g(x, y).$$

An example which is not of this type is the sphere of radius 5 about the origin, which is the level surface

$$x^2 + y^2 + z^2 = 25.$$

We wish to find the normal line and tangent plane to such a surface at a point (x_o, y_o, z_o) of the surface. If we can find a vector perpendicular, in some sense, to the surface at (x_o, y_o, z_o), then we shall be done. For this will be a vector A parallel to the normal line and perpendicular to the tangent plane. What we shall show is that grad f, evaluated at (x_0, y_0, z_0), is such a vector A.

But what is it to mean for a vector to be perpendicular to the surface at (x_0, y_0, z_0)? If we knew there were lines on the surface through (x_0, y_0, z_0), we might say that the vector should be perpendicular to all such lines. But there are no lines on the sphere, for instance; so we next try to find a vector perpendicular at (x_0, y_0, z_0) to all curves on the surface passing through (x_0, y_0, z_0). Therefore let $X(t) = (x(t), y(t), z(t))$ be a curve on the surface, and let $(x_0, y_0, z_0) = X(t_0)$. Since the curve lies on the surface, we know that

$$f(x(t), y(t), z(t)) = c \quad \text{for all} \quad t.$$

Therefore

$$\frac{d}{dt}(f(x(t), y(t), z(t))) = \frac{dc}{dt} = 0.$$

But we have seen that

$$\frac{d}{dt}(f(x(t), y(t), z(t))) = (\text{grad } f) \cdot X'(t),$$

so

$$(\text{grad } f) \cdot X'(t) = 0.$$

In particular,

$$((\text{grad } f)(x_0, y_0, z_0)) \cdot X'(t_0) = 0.$$

But $X'(t_0)$ is a vector parallel to the tangent to the curve $X(t)$ at $t = t_0$. Since $(\text{grad } f)(x_0, y_0, z_0)$ is perpendicular to $X'(t_0)$, it is perpendicular to the tangent to the curve at t_0, and this is true for any curve on the surface passing through (x_0, y_0, z_0). Thus $(\text{grad } f)(x_0, y_0, z_0)$ is the desired vector A. This gives a geometric interpretation to the direction of grad f: grad f, evaluated at (x_0, y_0, z_0), is perpendicular to the level surface $f(x, y, z) = f(x_0, y_0, z_0)$ at the point (x_0, y_0, z_0). We shall give an interpretation of its length in the next paragraph.

D. Directional derivative. Let $f(x, y, z)$ be a function of three variables (although n variables would do as well). Let $A = (a_1, a_2, a_3)$ be a given non-zero vector, and let $X_0 = (x_0, y_0, z_0)$ be a given point. When we find f_x, f_y, f_z at X_0, we are finding the rates of change of f with respect to distance traveled from the point X_0 in one of the three directions parallel to the respective axes. Now suppose instead that we locate the vector A with the point X_0 as initial point and travel along A, or in other words, we travel from X_0 in the same direction as A, and we wish to measure the rate of change of f with respect to the distance traveled. If we go from X_0 to $X_0 + dX$, the "average rate" will be

$$\frac{f(x_0+dx, y_0+dy, z_0+dz) - f(x_0, y_0, z_0)}{|dX|}.$$

Now the vector $dX = (X_0 + dX) - X_0$ is to have the same direction as A, so $dX = hA$ for some $h > 0$. Therefore $|dX| = h|A|$. Moreover,

$$f(x_0 + dx, y_0 + dy, z_0 + dz) - f(x_0, y_0, z_0)$$

$$= (\text{grad } f) \cdot dX + \text{error},$$

where

$$\frac{\text{error}}{|dX|} \longrightarrow 0 \quad \text{as} \quad |dX| \longrightarrow 0.$$

The "instantaneous rate", which we call the directional derivative of $f(x, y, z)$ at X_0 in the direction of A, will be the limit as $|dX| \longrightarrow 0$ of the average rate, or since $|dX| \longrightarrow 0$ means the same as $h \longrightarrow 0$,

$$\lim_{h \longrightarrow 0} \frac{(\text{grad } f) \cdot hA}{h|A|} + \frac{\text{error}}{h|A|}.$$

But we have seen that the second term goes to zero, and the first is simply

$$\frac{(\text{grad } f) \cdot A}{|A|}$$

This result gives us a simple formula for finding the directional derivative.

Now observe that if θ is the angle between A and grad f, then $(\text{grad } f) \cdot A = |\text{grad } f|\ |A| \cos \theta$, or the directional derivative is simply $|\text{grad } f| \cos \theta$. This is of course largest when $\cos \theta = 1$, i.e., when A has the same direction as grad f, and in that case it is just $|\text{grad } f|$. Thus we have our interpretation for $|\text{grad } f|$: It is the maximum value for the directional derivative of f in any direction. Note also that the minimum value for the directional derivative of f occurs when the direction is opposite to that of grad f, and that this minimum value is $-|\text{grad } f|$.

EXERCISES

1) Investigate the following functions for relative maxima and minima:

a) $f(x, y, z) = \sqrt{3 - x^2 - y^2 - z^2}$

b) $f(x, y) = \frac{1}{2}(x^2 + y^2)$

c) $f(x, y) = \frac{1}{2}(x^2 - y^2)$

2) Let $f(x, y, z) = z - e^x \sin y$, $P = (\log 3, \frac{3\pi}{2}, -3)$.

Find:
a) grad f at P;
b) the normal line at P to the level surface for f which passes through P;
c) the tangent plane to this surface at P;
d) the directional derivative of f at P in the direction of the vector $(1, 2, 2)$.
e) the maximum and minimum values for the directional derivative of f at P.

3) Consider the formula for kinetic energy: $K = \frac{1}{2}mv^2$. Suppose that m is found to be 20 grams, with a maximum error of $\pm$ .01 gram, and v is found to be 100 cm/sec, with a maximum error of $\pm$ 2 cm/sec. Estimate the maximum error resulting in the computation of the kinetic energy, as well as the percentage error in K.

4) Show that $f(x, y, z)$ is a constant if and only if grad $f = 0$.

23. The Real Number System

Now we shall return to first concepts and make precise some ideas which have been left somewhat indefinite. In the following sections we supply proofs of the basic theorems which we have used without proof. It will be seen along the way that the tools available to you from your previous background were totally inadequate for proving, say, that a function continuous on an interval has a maximum there.

To begin with, we set down explicitly the properties of the numbers which we use. There are several ways of doing this. One of them sets down the properties of the positive integers, and builds from them a system with the properties which we list below. Another, which we prefer for this course, takes the form of establishing a set of ground rules, or axioms: "The system of numbers with which we shall work is to have the properties ... ." The axioms fall in a natural fashion into three groups, which we call the field axioms, the order axioms, and the axiom of continuity. Throughout these axioms, the letters a, b, c, ... represent members of a set of objects which we agree to call numbers. In this set, we assume that we can add a to b to get a number $a + b$, and that we can <u>multiply</u> a by b to get a number ab. The first group of axioms, the <u>field axioms</u>, express properties of these operations from which all other properties can be derived.

<u>Group I. Field Axioms</u>

A1) $a + b = b + a$

A2) $(a + b) + c = a + (b + c)$

A3) There is exactly one number 0 such that $a + 0 = a$ for every a.

A4) For each a, there is exactly one number $-a$ such that $a + (-a) = 0$.

M1) $ab = ba$

M2) $(ab)c = a(bc)$

M3) There is exactly one number $1 \neq 0$ such that for every a, $a \cdot 1 = a$.

M4) If $a \neq 0$, there is exactly one number a^{-1} such that $aa^{-1} = 1$.

D) $a(b + c) = ab + ac$.

Axioms A1) - 4) give the properties of addition, M1) - 4) the properties of multiplication. Axiom D), the <u>distributive law</u>, connects the two operations. A1) and M1) express the <u>commutative laws</u> for addition and multiplication, respectively, A2) and M2) the <u>associative laws</u>. The number 0 of A3) is of course called "zero" and the 1 of M3) is called "one", although it is to be regarded as different from the "one" used in counting. The number $-a$ of A4) is called the <u>negative</u> of a, the number a^{-1} of M4) the <u>inverse</u> of a. One cannot define a system in which the field axioms are satisfied and which contains an inverse for 0. Observe that the field axioms are satisfied by a much larger class of systems than our previous notion of real numbers; for example, they are satisfied by the rational numbers.

We supply some sample proofs to illustrate our assertion that all properties of addition and multiplication follow from the field axioms.

THEOREM. $-(-a) = a$.

PROOF. $a + (-a) = 0$; $(-a) + a = a + (-a) = 0$, by A1); But $-(-a)$ is the <u>only</u> number b such that $-a + b = 0$, by A4). Therefore $a = -(-a)$.

THEOREM. $a \cdot 0 = 0$.

PROOF. $a = a \cdot 1$, by M3); but $1 = 1 + 0$, by A3). By D),

$a = a \cdot 1 = a \cdot (1 + 0) = a \cdot 1 + a \cdot 0 = a + a \cdot 0$. Therefore $(-a) + a = (-a) + (a + a \cdot 0) = ((-a) + a) + a \cdot 0$, by A2). But $(-a) + a = 0$, by A2) and A4), so we have $0 = 0 + a \cdot 0 = a \cdot 0$, by A3).

THEOREM. $(-1) \cdot a = -a$.

PROOF. $0 = 1 + (-1)$. Therefore $a \cdot 0 = a(1 + (-1)) = a \cdot 1 + a(-1)$, or by the last theorem, $0 = a \cdot 1 + a(-1) = a + (-1)a$, by M3) and M1). Since $-a$ is the only number b such that $a + b = 0$, we have $(-1)a = -a$.

THEOREM. $(-a)(-b) = ab$.

PROOF. $b + (-b) = 0$. Therefore $a(b + (-b)) = a \cdot 0 = 0$, or by D), $ab + a(-b) = 0$. Thus $a(-b) = -(ab)$. Also $a + (-a) = 0$, so $(a + (-a))(-b) = a(-b) + (-a)(-b) = 0$. Thus $a(-b) = -(-a)(-b)$, or $-(ab) = -(-a)(-b)$. Taking negatives once more, we see by the first theorem that $ab = (-a)(-b)$.

Observe that throughout these proofs we have used what you may have seen as the principle that "equals may be substituted for equals." But it has not been a case of substitution; the number 0 is the same whether it is written as 0, $a + (-a)$, $(-b) + b$, or what have you. It is only the representation of the number which changes, and we simply use the principle that if a rule holds for a number, it does not matter how we represent that number in applying the rule.

EXERCISE

Prove that $-(a + b) = (-a) + (-b)$; $-(b - a) = a - b$, where $a - b = a + (-b)$.

The second group of axioms deals with the ordering of numbers. We give them the form of axioms about positive numbers, and then define $a < b$ to mean "$b - a$ is positive."

Group II. Order Axioms.

We take as basic for all these axioms the existence of a certain set P of numbers, which we shall call "positive numbers." The axioms give properties that P is to have.

P1) If a and b are in P, so is $a + b$.

P2) If a and b are in P, so is ab.

P3) If a is any number $\neq 0$, then either a is in P or $-a$ is in P, but not both.

P4) 0 is not in P.

Now we define $a < b$ to mean "$b - a$ is in P". $a \leqq b$ is to mean "either $a < b$ or $a = b$". In place of $a < b$ we sometimes write $b > a$ and similarly $b \geqq a$ for $a \leqq b$. From the axioms of Groups I and II we can prove all the properties of $a < b$:

THEOREM. For any two numbers a and b, exactly one of the following holds: $a < b$, $b < a$, $a = b$.

PROOF. Consider the number $b - a$. If $b - a$ is in P, then $a < b$, $-(b - a) = a - b$ is not in P, and $a \neq b$. Then $a < b$ is the case, and neither of $a = b$, $b < a$ is the case. Next suppose $b - a$ is not in P. Then $a < b$ is not the case. But if $a \neq b$, $-(b - a) = a - b$ is in P, by P3); in other words, $b < a$. The only remaining possibility is $a = b$, in which case neither $b - a$ nor $a - b$ is in P, since both are 0.

THEOREM. If $a < b$, then $a + c < b + c$.

PROOF. Let $a < b$, i.e., $b - a$ is in P. But $(b + c) - (a + c) = b + c - a - c = b - a + c - c = b - a$ is in P. Therefore $a + c < b + c$.

THEOREM. If $a < b$ and $c > 0$ (i.e., c is in P), then $ca < cb$.

PROOF. $b - a$ is in P, and c is in P. Therefore $c(b - a) = cb - ca$ is in P, by P2). Thus $ca < cb$.

THEOREM. If $a < b$ and $b < c$, then $a < c$.

PROOF. If $a < b$ and $b < c$, then both $b - a$ and $c - b$ are in P. Then, by P1), so is $(c - b) + (b - a) = c - a$, or $a < c$.

THEOREM. If $a \neq 0$, then a^2 is in P.

PROOF. Either a is in P or $-a$ is in P. If a is in P, then by P2), $aa = a^2$ is in P. If $-a$ is in P, then $(-a)(-a) = a \cdot a = a^2$ is in P. This completes the proof.

COROLLARY. $1 > 0$, i.e., 1 is in P.

PROOF. $1 = 1^2$.

EXERCISES

Prove the following:

1. If $a < b$, then $-b < -a$.
2. If $a < b$ and $c < 0$, then $bc < ac$.
3. If $ab > 0$, then either both $a > 0$ and $b > 0$ of both $a < 0$ and $b < 0$.
4. If $a > 0$, then $a^{-1} > 0$; if $a < 0$, then $a^{-1} < 0$.
5. If $a < b$ and $c < d$, then $a + c < b + d$.
6. The integers are a system which satisfies all axioms of Groups I and II except M4). By considering pairs (m, n) (or $\frac{m}{n}$, if you wish) of integers with $n \neq 0$, develop the operations and ordering of the rational numbers. Note that the rational number which we represent by $\frac{2}{3}$ can also be represented by $\frac{6}{9}$, $\frac{4}{6}$, $\frac{-2}{-3}$, etc. Therefore $\frac{m}{n}$ will not itself be a rational number, but will be a way of representing a rational number. You should make it clear when two pairs $\frac{m}{n}$, $\frac{m'}{n'}$ represent the same rational number.

Group III. Axiom of Continuity.

All the axioms of Groups I and II are satisfied by the rational numbers. But within the rational numbers, we cannot prove the theorem which we have used in Section 12, namely that a continuous function $f(x)$ on $a \leq x \leq b$, such that $f(a) < f(b)$, takes on every value between $f(a)$ and $f(b)$ in the interval. For example, we consider the dilemma of the Greeks, who had no number to represent the length of the diagonal of a square of side 1. By the theorem of Pythagoras, the diagonal d would satisfy $d^2 = 1^2 + 1^2 = 2$. But there is no rational number d with this property; for if there were, we write $d = \frac{m}{n}$ as a fraction in lowest terms, where m and n are integers. Then

$$d^2 = \frac{m^2}{n^2} = 2,$$

or $m^2 = 2n^2$. Thus m^2 is divisible by 2. But if m is odd, this is impossible, therefore m is even, say $m = 2p$. Then $4p^2 = 2n^2$, or $n^2 = 2p^2$, and so n is also even. Therefore $\frac{m}{n}$ cannot have been in lowest terms, since we can cancel a 2. This is a contradiction, meaning that the original assumption that d was rational is false.

In terms of the intermediate value theorem, this can be interpreted as follows: $y = f(x) = x^2$ is continuous for $0 \leq x \leq 2$, and $f(0) = 0 < 4 = f(2)$. If the intermediate value theorem were to hold for rational numbers, there would be a rational number x_0 between 0 and 2 such that $x_0^2 = 2$. But there is no such rational number. In order to prove the intermediate value theorem we thus need a further property that a system satisfying only the field and order axioms need not possess. It is the fact that you had no experience with this property which made it impossible to prove the theorem at the time it was stated. Before we give the axiom, however, we must make a few definitions.

Let S be any set of numbers. A number M is called an upper bound for S if $x \leq M$ whenever x is in S. A number N is a lower bound for S if $N \leq x$ whenever x is in S. The set of all numbers has no upper or lower bounds, for instance, if M were to be an upper bound, then $M + 1$ would be a number and $M + 1 > M$. The set of all negative numbers has many upper bounds, for instance 41, $29\frac{3}{8}$, 0, π, $\sqrt{2}$, but no lower bound.

A number M is called a least upper bound for S if:

a) M is an upper bound for S; and

b) there is no upper bound M' for S such that $M' < M$.

Greatest lower bound is similarly defined. A least upper bound for the set of negative numbers is 0; this is also a least upper bound for the set of non-positive numbers. If S contains no numbers, then any number is an upper bound for S, so S cannot have a least upper bound. It is evident that if S has a least upper bound, it has only one. Now we can state our final axiom:

C) If a non-empty set S has an upper bound, it has a least upper bound.

It follows that if S has a lower bound, it has a greatest lower bound. For if $-S$ denotes the set of negatives of the elements of S, and if M is a lower bound for S, then $-M$ is an upper bound for $-S$. Therefore $-S$ has a least upper bound M', and $-M'$ is a greatest lower bound for S. As a first application, we prove that the integers have no upper bound.

THEOREM. There is no upper bound for the set 1, 2, 3, ...

PROOF. Suppose the theorem is false; then by C), the integers have a least upper bound M. Then $M - 1$ cannot be an upper bound, so there is an integer $n > M - 1$. But then $n + 1 > (M - 1) + 1 = M$, and $n + 1$ is an integer. This contradicts the assumption that M was an upper bound, and so the theorem is proved. (Note that it follows that if ϵ is any positive number, there is a positive integer n such that $\frac{1}{n} < \epsilon$; for otherwise, $\frac{1}{\epsilon}$ would be an upper bound for the positive integers.)

Now we know that there is in our system a number d such that $d^2 = 2$. This will follow from the continuity of x^2 and the intermediate value theorem, but let us give an indication of how it could be shown directly. Let S be the set of positive rational numbers r such that $r^2 < 2$. Then 1 is in S, so that S is non-empty. 2 is an upper bound for S, for if $r > 2$, then $r^2 > 2^2 = 4 > 2$, and r is not in S. So let d be the least upper bound for S. If $d^2 > 2$, let $d^2 - 2 = \epsilon > 0$. Let n be an integer such that $\frac{4}{n} < \epsilon$, or $\frac{1}{n} < \frac{\epsilon}{4}$. Then

$$\left(d - \frac{1}{n}\right)^2 = d^2 - \frac{2d}{n} + \frac{1}{n^2} \geqq d^2 - \frac{4}{n} + \frac{1}{n^2},$$

since $d \leqq 2$, or

$$\left(d - \frac{1}{n}\right)^2 \geqq d^2 - \epsilon + \frac{1}{n^2} > d^2 - \epsilon = 2.$$

Thus $d - \frac{1}{n}$ is an upper bound for S and is less than d, which is impossible. Therefore $d^2 \leqq 2$. If $d^2 < 2$, we find a rational number r in S such that $r > d$. (If $2 - d^2 = \delta > 0$, let n be a positive integer such that $\frac{1}{n^2} < \delta$; then let q be a positive integer so large that there is a square of an integer, m, between $(2n^2 - 1)q^2$ and $2n^2q^2$. Let $r = \frac{m}{qn}$. Then $r^2 < 2$, and $r > d$.) The only remaining possibility is $d^2 = 2$.

24. Sequences and Limits

A sequence of numbers is really a function which assigns to each positive integer a number. We denote it customarily by $\{a_n\}$, where $a_n = f(n)$ for each integer $n > 0$. Examples are:

a) $1, -1, 1, -1, 1, -1, \ldots$;

$a_n = 1$ if n is odd, -1 if n is even.

b) $1, \frac{1}{2}, \frac{1}{3}, \frac{1}{4}, \ldots$;

$a_n = \frac{1}{n}$.

c) $1, 1 + 1, 1 + 1 + \frac{1}{2!}, \ldots$;

$a_n = 1 + \frac{1}{1!} + \frac{1}{2!} + \cdots + \frac{1}{(n-1)!}$.

The sequence $\{a_n\}$ is said to have the limit a if for every given positive number ϵ (usually considered as small), we can find an integer N such that whenever $n > N$, we have $a - \epsilon < a_n < a + \epsilon$, or, what is the same, $|a - a_n| < \epsilon$. That is to say, I can only assert that the limit is a if every opponent who chooses to challenge me on the matter can be dealt with; I must be prepared to give a procedure for finding an N with the desired property on the basis of any ϵ with which he may wish to confront me. If $\{a_n\}$ has the limit a, we write either

$$\lim_{n \to \infty} a_n = a, \quad \text{or simply} \quad \lim a_n = a.$$

The former is read, "The limit as n tends to infinity of a_n is a". Note that we have nowhere defined any object called "infinity", but rather we have chosen this whole idiom to refer to the behavior which has been given a precise definition above.

The sequence a) has no limit (try to meet the challenge of $\epsilon = \frac{1}{2}$). The sequence b) has the limit 0, and the sequence c) has the limit e, as we have seen in Section 17. Suppose however that we are given a sequence $\{a_n\}$ and it is not evident what its limit is. Can we tell whether it has a limit without first guessing the limit? The answer to this is yes, by means of several possible devices. The most powerful one, the Cauchy convergence criterion, will be given with its proof as an appendix to this section. One which is somewhat simpler may be applied if the sequence is increasing: $a_1 \leqq a_2 \leqq a_3 = \ldots$.

THEOREM. Let $\{a_n\}$ be increasing, and let $\{a_n\}$ have an upper bound. Then $\lim a_n = a$ for some number a.

PROOF. Let a be the least upper bound for $\{a_n\}$. Let $\epsilon > 0$. We must show that there is an N such that whenever $n > N$, $a - \epsilon < a_n < a + \epsilon$. But $a_n < a + \epsilon$ for all n, since $a_n \leqq a$ for all n (a is an upper bound). Now $a - \epsilon$ is not an upper bound, so there is an N such that $a_N > a - \epsilon$. If $n > N$, then $a_n \geqq a_N > a - \epsilon$. Thus for $n > N$, we have $a - \epsilon < a_n < a + \epsilon$, which was to be proved. (Note that N will in general depend on ϵ.)

If we did not know that the limit of the sequence c) is e, we could show by this theorem that it has a limit. For the sequence is obviously increasing, and

$$1 + 1 + \frac{1}{2!} + \frac{1}{3!} + \cdots + \frac{1}{(n-1)!} \leqq 1 + 1 + \frac{1}{2} + \frac{1}{2^2} + \cdots + \frac{1}{2^{n-2}}$$

$$= 1 + \frac{1 - (\frac{1}{2})^{n-1}}{1 - \frac{1}{2}} < 3 \quad \text{for all } n.$$

Therefore we have a bounded increasing sequence. The theorem can also be applied, using negatives, to show that if a decreasing sequence has a lower bound, it has a limit. (Note also that two numbers a and b cannot both be limits of $\{a_n\}$; for then we could not meet the challenge of $\epsilon = \frac{1}{2}|b - a|$ for either point.)

As an example of a sequence whose limit we can not guess, let $a_n = 1 + \frac{1}{2} + \dots + \frac{1}{n} - \log n$. Then

$$a_{n+1} - a_n = \frac{1}{n+1} - \log(n+1) + \log n = \frac{1}{n+1} - \log\left(1 + \frac{1}{n}\right).$$

But if we recall from Section 11 the inequality

$$\log\left(1 + \frac{1}{n}\right) \geqq \frac{1}{n+1}, \quad \text{we see that} \quad a_n \geqq a_{n+1},$$

or that <u>the sequence is decreasing</u>. Now let us show that it has 0 as a lower bound. To do this, recall the other side of the same inequality on $\log x$:

$$\frac{1}{n} \geqq \log\left(1 + \frac{1}{n}\right) = \log(n+1) - \log n .$$

Thus

$$a_n \geqq (\log 2 - \log 1) + (\log 3 - \log 2) + \dots + (\log(n+1) - \log n) - \log n$$

$$= (\log 2 - \log 2) + (\log 3 - \log 3) + \dots + (\log n - \log n) + \log(n+1)$$

$$- \log n = \log(n+1) - \log n \geqq 0 \quad \text{for all } n.$$

Therefore

$$\lim_{n \to \infty} \left(1 + \frac{1}{2} + \dots + \frac{1}{n} - \log n\right) = C$$

exists, and is in fact non-negative. Its first few decimals are .5772 ... . It is called <u>Euler's constant</u>, and it is not even known whether it is rational.

Next we prove some properties of limits of sequences, including the <u>linearity</u> of the limits. Some others are left as exercises. In general, let $\{a_n\}$ be a sequence with limit a, $\{b_n\}$ a sequence with limit b, and c a constant.

THEOREM. $\lim (a_n + b_n) = a + b$.

PROOF. Let $\epsilon > 0$ be given. Then we can find N_1 such that for all $n > N_1$, $-\epsilon < a_n - a < \epsilon$, and N_2 such that for all $n > N_2$, $-\epsilon < b_n - b < \epsilon$. Let N be the <u>larger</u> of N_1 and N_2, written $N = \text{Max}(N_1, N_2)$. Then for all $n > N$, $(a_n + b_n) - (a + b)$ has the property that

$$-2\epsilon < (a_n + b_n) - (a + b) < 2\epsilon .$$

The factor of 2 makes no difference, since we could have started with $\frac{\epsilon}{2}$ instead of ϵ and arrived at $-\epsilon < (a_n + b_n) - (a + b) < \epsilon$ for all $n > N'$ for a suitable N'. Therefore the theorem is proved.

THEOREM. $\lim(ca_n) = ca$.

PROOF. If $c = 0$, then all $ca_n = ca = 0$, and for any given $\epsilon > 0$, we can take $N = 1$. Then $-\epsilon < ca_n - ca = 0 < \epsilon$, for all $n > N$. Now let $c \neq 0$, and let $\epsilon > 0$ be given. Let N_1 be such that $-\epsilon < a_n - a < \epsilon$ for all $n > N_1$, or in other words, $|a_n - a| < \epsilon$ for $n > N_1$. Then

$$|ca_n - ca| = |c|\ |a_n - a| < |c|\ \epsilon \quad \text{for} \quad n > N_1 .$$

We need only start with $\frac{\epsilon}{|c|}$ instead of ϵ to get a result of the form

$$|ca_n - ca| < \epsilon \quad \text{for} \quad n > N.$$

THEOREM. $\lim (a_n b_n) = ab$.

PROOF IN PART.

$$|a_n b_n - ab| = |a_n b_n - a_n b + a_n b - ab|$$

$$\leqq |a_n b_n - a_n b| + |a_n b - ab| = |a_n|\ |b_n - b| + |a_n - a|\ |b|.$$

Now consider $\{a_n\}$, with $\epsilon = 1$. There is an N_1 such that for $n > N_1$, $|a_n - a| < 1$. Now we claim that $|a_n| - |a| \leqq |a_n - a|$. For

$$|a_n| = |(a_n - a) + a| \leqq |a_n - a| + |a|,$$

or

$$|a_n| - |a| \leqq |a_n - a|.$$

Thus we have $|a_n| - |a| \leqq |a_n - a| < 1$, or $|a_n| < 1 + |a|$ for all $n > N_1$. Therefore if $n > N_1$,

$$|a_n b_n - ab| \leqq (|a| + 1)\ |b_n - b| + |a_n - a|\ |b|.$$

The remainder of the proof is left as an exercise.

EXERCISE

Suppose $\lim a_n = a$, $a \neq 0$.

1) Show that there is a positive number r and an integer N_1 such that for all $n > N_1$, $|a_n| > r$.

2) Now show that $\lim \left(\frac{1}{a_n}\right) = \frac{1}{a}$, in the sense that for any given $\epsilon > 0$ there is an N such that for all $n > N$, $a_n \neq 0$ and $|\frac{1}{a_n} - \frac{1}{a}| < \epsilon$.

Next let $f(x)$ be a function defined on an interval containing a point x_0, with the possible exception that $f(x_0)$ may not be defined. We say that a is the <u>limit as</u> x <u>approaches</u> x_0 of $f(x)$, written

$$\lim_{x \to x_0} f(x) = a,$$

if for every given $\epsilon > 0$ we can find a $\delta > 0$ such that for every x in the interval $0 < |x - x_0| < \delta$ we have $|f(x) - a| < \epsilon$.

EXERCISES

Let

$$\lim_{x \to x_0} f(x) = a, \qquad \lim_{x \to x_0} g(x) = b,$$

c a constant. Prove:

1) $\lim_{x \to x_0} (f(x) + g(x)) = a + b.$

2) $\lim_{x \to x_0} (cf(x)) = ca.$

3) $\lim_{x \to x_0} (f(x)g(x)) = ab.$

4) If $a \neq 0$, $\lim_{x \to x_0} \left(\frac{1}{f(x)}\right) = \frac{1}{a}$,

5) $\lim_{x \to x_0} f(x) = a$ means the same as $\lim_{h \to 0} f(x_0 + h) = a.$

APPENDIX: CAUCHY'S CRITERION

As a variation, we prove this theorem for the limit of a <u>function</u>, and leave its proof for <u>sequences</u> as an exercise.

THEOREM. Let $f(x)$ be defined on an interval

containing x_o, but not necessarily <u>at</u> x_o. Then

$$\lim_{x \to x_o} f(x)$$

exists if and only if $f(x)$ satisfies the following condition: For every $\epsilon > 0$ there is a $\delta > 0$ such that if x_1 and x_2 are any numbers with $0 < |x_1 - x_o| < \delta$, $0 < |x_2 - x_o| < \delta$, then $|f(x_1) - f(x_2)| < \epsilon$.

PROOF. First suppose the limit exists, say

$$\lim_{x \to x_o} f(x) = a.$$

Let $\epsilon > 0$ be given. Then there is a $\delta > 0$ such that if $0 < |x - x_o| < \delta$, we have $|f(x) - a| < \frac{\epsilon}{2}$. Let $0 < |x_1 - x_o| < \delta$, $0 < |x_2 - x_o| < \delta$. Then

$$|f(x_1) - f(x_2)| = |f(x_1) - a + a - f(x_2)|$$

$$\leqq |f(x_1) - a| + |f(x_2) - a| < \frac{\epsilon}{2} + \frac{\epsilon}{2} = \epsilon.$$

Thus the theorem is proved in one direction: if the limit exists, the condition is satisfied.

Now suppose the condition is satisfied. First take $\epsilon = 1$; then let $0 < |x_1 - x_o| < \delta_o$, where δ_o satisfies the condition for $\epsilon = 1$. Now if x is any number such that $0 < |x - x_o| < \delta_o$, we have $|f(x) - f(x_1)| < 1$, or $|f(x)| < |f(x_1)| + 1$, i.e., $f(x)$ <u>is bounded</u> from above and from below for $0 < |x - x_o| < \delta_o$.

Next let $\delta_1 = \text{Min}\,(1, \delta_o)$, $\delta_2 = \text{Min}\,(\frac{1}{2}, \delta_o)$, ..., $\delta_n = \text{Min}\,(\frac{1}{n}, \delta_o)$. (From some n on, all $\delta_n = \frac{1}{n}$, since eventually $\frac{1}{n} < \delta_o$.) On each of the intervals $I_n : 0 < |x - x_o| < \delta_n$, $f(x)$ is bounded from above and from below. Let a_n be <u>the least upper bound of the values of</u> $f(x)$ <u>on</u> I_n. The sequence $\{a_n\}$ is <u>decreasing</u>, since a least upper bound on I_n is an upper bound on I_{n+1}. Moreover, all $a_n \geqq L$, where L is any lower bound for $f(x)$ on I_1. Therefore $\lim a_n$ exists. Let us call it a. Now we shall prove that

$$a = \lim_{x \to x_o} f(x).$$

Let $\epsilon > 0$ be given. Then there is an N such that for all $n > N$, $|a - a_n| < \frac{\epsilon}{3}$. Since a_n is the <u>least</u> upper bound for $f(x)$ in I_n, there is an x_n in I_n, such that $|a_n - f(x_n)| < \frac{\epsilon}{3}$, and this for each n. There is also a δ' such that whenever $0 < |x - x_0| < \delta'$, $0 < |x' - x_0| < \delta'$, then $|f(x) - f(x')| < \frac{\epsilon}{3}$. Now take $n > N$ such that $\delta_n < \delta'$. We assert that δ_n is a suitable δ for the definition of limit. That is, let $0 < |x - x_0| < \delta_n$. Then $0 < |x - x_0| < \delta'$, and also $0 < |x_n - x_0| < \delta_n < \delta'$. Therefore

$$|f(x) - f(x_n)| < \frac{\epsilon}{3}.$$

Now we know that $|f(x_n) - a_n| < \frac{\epsilon}{3}$, and since $n > N$,

$$|a_n - a| < \frac{\epsilon}{3}.$$

Therefore

$$\begin{aligned} |f(x) - a| &= |f(x) - f(x_n) + f(x_n) - a_n + a_n - a| \\ &\leqq |f(x) - f(x_n)| + |f(x_n) - a_n| + |a_n - a| \\ &< \frac{\epsilon}{3} + \frac{\epsilon}{3} + \frac{\epsilon}{3} = \epsilon. \end{aligned}$$

We see that whenever $0 < |x - x_0| < \delta_n$, we have $|f(x) - a| < \epsilon$. Therefore

$$\lim_{x \to x_0} f(x) = a,$$

and the proof is complete.

EXERCISE

Prove Cauchy's criterion for sequences;

$\lim a_n$ exists if and only if the following condition holds:

For any $\epsilon > 0$ there is an N such that whenever $m > N$ and $n > N$, then $|a_m - a_n| < \epsilon$.

25. Continuous Functions

Let $f(x)$ be defined on an interval containing x_0, and assume now that $f(x_0)$ is defined. Then we say that $f(x)$ is continuous at x_0 if

$$\lim_{x \to x_0} f(x) = f(x_0),$$

or in other words, if for every $\epsilon > 0$ we can find a $\delta > 0$ such that whenever $|x - x_0| < \delta$, then $|f(x) - f(x_0)| < \epsilon$.

This is the precise way of saying "$f(x)$ is near $f(x_0)$ when x is near x_0", or that "a small change in x produces only a small change in $f(x)$." Note, however, that it is not the change in x that we start with, but rather the number ϵ, which is to be a bound for the change in $f(x)$. Then we find an interval about x_0, of length 2δ, such that if x is in this interval, i.e., x is not farther than δ away from x_0, then $f(x)$ is not farther away from $f(x_0)$ than the given bound ϵ. It is only by this type of definition that we can be sure of meeting the challenge of a man who says, "I don't call $f(x)$ near $f(x_0)$ until it is within 10^{-4}", a second one who says "Near means to me within 10^{-8}," and so on. Thus in order to show that a given function $f(x)$ is continuous at x_0, we must start with a symbol ϵ, representing a general positive number, and then show that we can find a positive number δ, which will usually depend on ϵ, such that the inequality $|x - x_0| < \delta$ gives us some information about the expression $|f(x) - f(x_0)|$, namely that it is less than ϵ. To prove that $f(x)$ is not continuous at x_0, it would suffice to demonstrate just one ϵ for which we could not find a suitable δ. (Of course, then we could not find a suitable δ for any smaller ϵ.)

Let us illustrate by discussing some examples:

1) $f(x) = x^2$, $x_0 = 1$. If you recheck our proof that a differentiable function is continuous, you will see that it is completely rigorous in terms of the definitions and the theorems about limits which we have now given. However, let us prove that x^2 is continuous at 1 using only the definition of continuity.

Let $\epsilon > 0$ be given. Then we must study $|f(x) - f(x_0)| = |x^2 - 1|$. Let us write $h = x - 1$, $x = 1 + h$. Then

$$|f(x) - f(x_0)| = |1 + 2h + h^2 - 1| = |2h + h^2| = |h|\ |2 + h|.$$

Now let us take $\delta = \text{Min}\ (1, \frac{\epsilon}{3})$. Then if $|h| < \delta$,

$$|2 + h| \leqq 2 + |h| < 3, \quad \text{and} \quad |h|\ |2 + h| < 3\ |h| < \epsilon.$$

Since $h = x - 1 = x - x_0$, we see that whenever $|x - x_0| < \delta$, we have $|f(x) - f(x_0)| < \epsilon$, and thus that $f(x)$ is continuous at x_0.

2) Our second example will seem somewhat more bizarre. Let $f(x)$ be defined for $0 \leqq x \leqq 1$ as follows:

If x is irrational, $f(x) = 0$.

If $x = \frac{p}{q}$ in lowest terms ($q > 0$, $p \geqq 0$, p and q integers), then

$$f(x) = \frac{1}{q}\,.$$

For example,

$$f\left(\frac{\sqrt{2}}{2}\right) = 0 = f\left(\frac{\pi}{4}\right) = f\left(\frac{e}{7}\right);\ f\left(\frac{1}{3}\right) = \frac{1}{3} = f\left(\frac{2}{3}\right);$$

$$f(0) = 1 \quad \left(0 = \frac{0}{1} \text{ in lowest terms}\right),\ f(1) = 1.$$

First of all, $f(x)$ is <u>not</u> continuous at x_0 if x_0 is a <u>rational</u> number; for if $x_0 = \frac{p}{q}$ in lowest terms, $f(x_0) = \frac{1}{q} > 0$. In every interval about x_0 we have irrational numbers x, where $f(x) = 0$. Therefore if we tried $\epsilon = \frac{1}{2q}$, we could <u>not</u> find a suitable δ. Now we show that $f(x)$ <u>is</u> continuous at x_0 if x_0 is any <u>irrational</u> number; then $f(x_0) = 0$. Let $\epsilon > 0$ be given, and find an integer $n > 0$ such that $\frac{1}{n} < \epsilon$. Now there are in the interval $0 \leqq x \leqq 1$ only a finite number of rational numbers which, in lowest terms, have denominators less than n; there are two with denominator 1, one with denominator 2, two with denominator 3, two with denominator 4, four with denominator 5, ... In general, except for denominator 1, there <u>at most</u> $(k - 1)$ with denominator k (and this is only achieved if k is prime). Therefore those with denominator <u>less than</u> n are at most

$$2 + 1 + 2 + \dots + (n - 2) = 2 + \frac{(n-1)(n-2)}{2}$$

in number. Some one of these, say r, is therefore closest to x_0. Let $\delta = |r - x_0|$. $\delta > 0$, since x_0 is irrational. Now let $|x - x_0| < \delta$. If x is <u>irrational</u>, then $|f(x) - f(x_0)| = |0 - 0| = 0 < \epsilon$. If $x = \frac{p}{q}$ is <u>rational</u>, we know that $q \geqq n$, or $f(x) = \frac{1}{q} \leqq \frac{1}{n} < \epsilon$. Thus $|f(x) - f(x_0)| = \frac{1}{q} < \epsilon$, and we have proved that $f(x)$ is continuous at x_0.

EXERCISES

1) Prove that $|x|$ is continuous at $x = 0$.
2) Prove that $x^{2/3}$ is continuous at $x = 0$.
3) Prove that

$$f(x) = \begin{cases} x \sin \frac{1}{x}, & x \neq 0 \\ 0, & x = 0 \end{cases} \text{ is continuous at } x = 0.$$

4) Prove that $[x]$ is continuous at x_0 if x_0 is <u>not</u> an integer, and that it is <u>not</u> continuous at x_0 if x_0 <u>is</u> an integer.

5) Prove the following, assuming that $f(x)$ and $g(x)$ are continuous at x_0:

a) $f(x) + g(x)$ is continuous at x_0.

b) $cf(x)$ is continuous at x_0, where c is a constant.

c) $f(x)g(x)$ is continuous at x_0.

d) If $f(x_0) \neq 0$, $\frac{1}{f(x)}$ is continuous at x_0.

6) Write out in formal language the proof that if $f'(x_0)$ exists, then $f(x)$ is continuous at x_0.

26. Theorems on Continuous Functions

Now let us supply proofs for the two theorems on continuous functions which we have used without proof. The use of the axiom of continuity in proving these should be noted carefully.

THEOREM. (Intermediate value theorem). Let $f(x)$ be continuous at every x, $a \leqq x \leqq b$, and let $f(a) \leqq y_0 \leqq f(b)$. Then there is an x_0, $a \leqq x_0 \leqq b$, such that $f(x_0) = y_0$.

PROOF. Let S be the set of all x such that $a \leqq x \leqq b$ and $f(x) \leqq y_0$. Then a is in S, so that S is not empty. Also, b is an upper bound for S. Therefore S has a least upper bound x_0. In the case of the function graphed below, x_0 is as indicated:

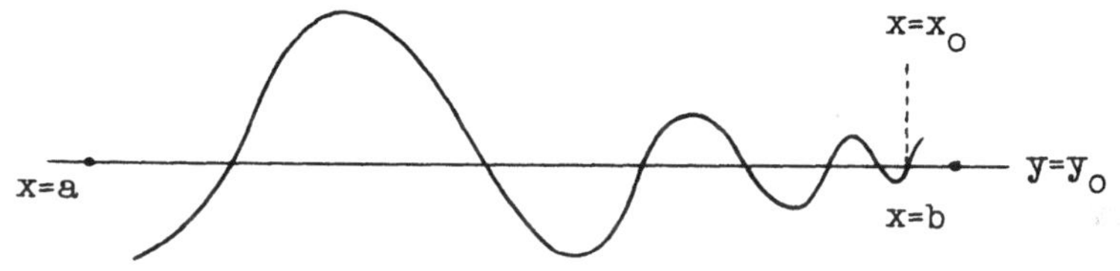

Now either $f(x_0) < y_0$, $f(x_0) > y_0$, or $f(x_0) = y_0$. We show that $f(x_0) = y_0$ by showing that the other two are impossible.

First of all, suppose $f(x_0) < y_0$. Let $\epsilon = y_0 - f(x_0) > 0$. Then we can find a $\delta > 0$ such that whenever $|x - x_0| < \delta$, we have $|f(x) - f(x_0)| < \epsilon$. Now we must have $x_0 \neq b$, since $f(b) \geqq y_0$. Therefore $h = \text{Min}\ (b - x_0, \frac{\delta}{2})$ is positive, and if we set $x = x_0 + h$, $|x - x_0| < \delta$, and

$$f(x) - f(x_0) \leqq |f(x) - f(x_0)| < \epsilon = y_0 - f(x_0), \quad \text{or} \quad f(x) < y_0.$$

Thus $x = x_0 + h$ is greater than x_0 and is in S. But x_0 was an upper bound for S. This is a contradiction, and therefore $f(x_0) \geqq y_0$. Here we have used only the property that x_0 is an upper bound for S.

Now suppose $f(x_0) > y_0$. We shall show that this is impossible by the fact that x_0 is the least upper bound for S. Clearly, $x_0 \neq a$. Let $\epsilon = f(x_0) - y_0$. Then $\epsilon > 0$, and we can find a $\delta > 0$ such that whenever $|x - x_0| < \delta$, then $|f(x_0) - f(x)| < \epsilon$, so

$$f(x_0) - f(x) \leqq |f(x_0) - f(x)| < \epsilon = f(x_0) - y_0,$$

$$-f(x) < -y_0, \quad \text{or} \quad f(x) > y_0 .$$

Therefore no values of x between $x_0 - \frac{\delta}{2}$ and x_0 are in S, and since no values of $x \geqq x_0$ are in S, no values of $x \geqq x_0 - \frac{\delta}{2}$ are in S. If $h = \text{Min}\,(x_0 - a, \frac{\delta}{2})$, then $x_0 - h$ is an upper bound for S, and $x_0 - h < x_0$. This contradicts the fact that x_0 was the least upper bound. Therefore $f(x_0) > y_0$ is impossible, and we must have $f(x_0) = y_0$. This completes the proof.

THEOREM. Let $f(x)$ be continuous for all x, $a \leqq x \leqq b$. Then $f(x)$ has a maximum on $a \leqq x \leqq b$, i.e., there is a number x_0, $a \leqq x_0 \leqq b$, such that $f(x) \leqq f(x_0)$ whenever $a \leqq x \leqq b$.

PROOF. Let S be the set of all x in the interval $a \leqq x \leqq b$ such that for some $x' > x$, $f(x') > f(x'')$ whenever $a \leqq x'' \leqq x$. In the diagram below, all points to the left of, and not including x_0 are in S: Since b is an upper bound for S, S has a least upper bound unless S is empty. But if S is empty, then a is not in S; i.e., for every $x' > a$, we have $f(x') \leqq f(a)$; this is simply the case where a is a maximum for $f(x)$, and in this case we are done. Thus there remains the case where S is not empty, and therefore has a least upper bound x_0. We shall prove that x_0 is a maximum for $f(x)$ on $a \leqq x \leqq b$, as in the diagram.

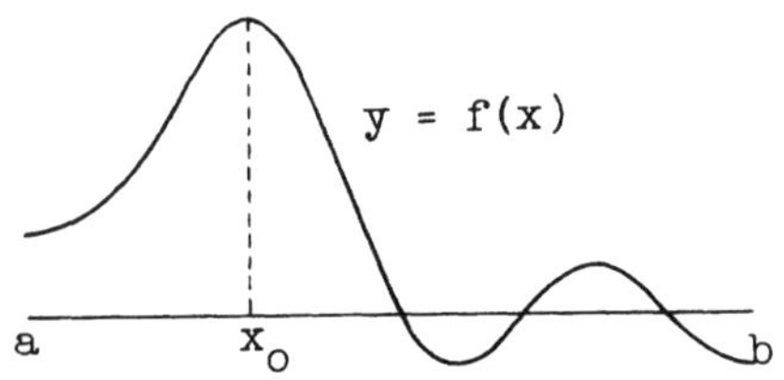

First we claim that x_0 is not in S. If $x_0 = b$, this is clear, since there is no x' of the interval such that $x' > b$. Thus we may assume $x_0 \neq b$. If x_0 were in S, we could find $x' > x_0$ such that $f(x') > f(x)$ whenever $a \leqq x \leqq x_0$. In particular, $f(x') > f(x_0)$. Then we could find a $\delta > 0$ such that whenever $|x - x_0| < \delta$, $f(x) - f(x_0) < f(x') - f(x_0)$, or $f(x) < f(x')$, by using $\epsilon = f(x') - f(x_0)$ in the continuity of $f(x)$ at x_0. We may also assume $\delta < b - x_0$. Then consider $x_0 + \frac{\delta}{2}$. For all x such that $a \leqq x \leqq x_0 + \frac{\delta}{2}$, we have $f(x) < f(x')$, or $x_0 + \frac{\delta}{2}$ is in S and is greater than x_0. This contradicts the fact that x_0 is an upper bound for S. Therefore x_0 is not in S.

Now let x be any point of the interval $a \leqq x \leqq b$. We shall show $f(x) \leqq f(x_0)$. Suppose first that $x > x_0$. If $f(x) > f(x_0)$, there is a $\delta > 0$ such that $f(x') < f(x)$ whenever $|x' - x_0| < \delta$, as before. By the fact that x_0 is the least upper bound, there is a number x_1 in S such that $x_0 - x_1 < \delta$. Consequently there is an $x_2 > x_1$ such that $f(x_2) > f(x')$ whenever $a \leqq x' \leqq x_1$. Either there is such an x_2 with $|x_2 - x_0| < \delta$ or there is not. If we do have one with $|x_2 - x_0| < \delta$, then $f(x) > f(x_2)$, and so $f(x) > f(x')$ whenever $a \leqq x' \leqq x_0$, or x_0 is in S. But this is impossible. Therefore we may assume $x_2 \geqq x_0 + \delta$. Then if $f(x) \geqq f(x_2)$, let $x_3 = x$; if $f(x) < f(x_2)$, let $x_3 = x_2$. In either case, $f(x_3) > f(x')$ whenever $a \leqq x' \leqq x_0$, so x_0 is in S, a contradiction. Thus we conclude that we must not have $f(x) > f(x_0)$, and therefore that $f(x) \leqq f(x_0)$ whenever $x \geqq x_0$.

Finally, we show that $f(x) \leqq f(x_0)$ when $a \leqq x \leqq x_0$. So assume that $a \leqq x < x_0$ and that $f(x) > f(x_0)$. Then there is an interval $|x' - x_0| < \delta$ where $f(x') < f(x)$. But there is a point x_1 of S such that $x_1 > x_0 - \delta$. Therefore there is an $x_2 > x_1$ such that $f(x_2) > f(x'')$ whenever $a \leqq x'' < x_1$; in particular, $f(x_2) > f(x)$. Therefore $x_2 \geqq x_0 + \delta$, or $x_2 > x_0$, and we have $f(x_2) > f(x) > f(x_0)$. But we have shown that this is impossible. Therefore $f(x) \leqq f(x_0)$, x_0 is a maximum for $f(x)$, and the theorem is proved.

EXERCISE

Let $f(x)$ be continuous for $a \leqq x \leqq b$. Let S be the set of all x, $a \leqq x \leqq b$, such that for some M, $f(x') \leqq M$ whenever $a \leqq x' \leqq x$. Prove that S has a least upper bound, and that this least upper bound is b. Note also that b is in S, therefore that the values of $f(x)$ have an upper bound for $a \leqq x \leqq b$. Let N be their least upper bound. If $f(x_0) = N$ for some x_0, then it is clear that x_0 is a maximum for $f(x)$. If $f(x) < N$ for all x of the interval, the function $\frac{1}{N - f(x)}$ is continuous for $a \leqq x \leqq b$, therefore has an upper bound B by the above. Show that this is impossible, and thus give another proof that $f(x)$ has a maximum for $a \leqq x \leqq b$.

In the year following that on which these notes are based, there was enough time at the end of the course to permit a rigorous discussion of the definite integral. It was pointed out that the concept of area used in our previous definition was only intuitive, and the problem of defining the concept of area under a more or less smooth curve was discussed. This led naturally to approximations above and below by rectangles, or to upper and lower sums. As in Begle, Introductory Calculus, (Holt, New York, 1954), the upper and lower integrals were defined as greatest lower bound and least upper bound, respectively, of the upper and lower sums, and the integral was defined as the common value of the upper and lower integrals when they are equal. The theorems about integrals previously used and the existence theorem for the integral of a continuous function were then proved as in §§3 and 4 of Chapter 7 of the book by Begle cited above.

Recommended reading and problems in applications for students of engineering. (Numbers at the left refer to sections in the course outline of calculus.)

6. Related rate problems.
   Reading: Oakley, pp. 58-59
   Problems: Oakley, p. 59.

   Arc length; curvature; evolutes.
   Reading: Oakley, pp. 67-74.
   Problems: Oakley, pp• 73-74.

9. Maxima and minima.
   Reading: Courant, pp. 163-166.
   Problems: Courant, pp. 166-167;
   p. 557, problems 70-73.

13. Hyperbolic functions.
    Reading: Courant, pp. 183-189.

14. Areas; arc length.
    Reading: Oakley, pp. 107-113
    Problems: Oakley, pp. 112-113.

    Volumes of revolution.
    Reading: Oakley, pp. 115-118
    Problems: Oakley, pp. 117-118.

    Work; force; center of mass; moment of inertia.
    Reading: Oakley, pp. 122-137.
    Problems: Oakley, pp. 124, 126, 132, 137.

    Distribution functions.
    Reading: Courant, pp. 122-126.

Differential equations: radioactive decay; temperature and pressure; growth and, decay in electric circuits.
Reading: Courant, pp. 178-183.
Problem: Courant, p. 183, problem 3.

14. Mechanics and work.
Reading: Courant, pp. 292-307.
Problems: Courant, p. 298; p. 570, problem 222.

Common problems of differential equations
Reading: Courant, pp. 502-509.
Problems: Courant, p. 509; p. 570, problem 220.

Printed in Dunstable, United Kingdom

91239726R00083